Praise For

THE ANSWER TO THE RIDDLE IS ME

"Incandescent . . . [A] profound and finely nuanced meditation on memory and identity."
—*Seattle Times*

"What does it mean to be the person you are? How much can be stripped away before you are no longer you? This is a fascinating book that resides in the mind as if you lived it yourself."
—Robert Boswell, author of *Tumbledown*

"[MacLean] writes eloquently about the bizarre and disturbing experience of having his sense of self erased and then reconstructed from scratch."
—*The New Yorker*

"Thoughtful, terribly honest, often funny, and utterly un-self-indulgent, this is a riveting work of narrative art."
—Tony Hoagland, author of *What Narcissism Means to Me*

"David Stuart MacLean is a writer who can break your heart, terrify you, and make you laugh all on the same page. *The Answer to the Riddle Is Me* is a masterful exploration of the funhouse of identity."
—Mat Johnson, author of *Pym*

"Brilliant and painful and hilarious."
—Antonya Nelson, author of *Some Fun*

"While MacLean's experience is unlucky indeed, the luck becomes ours as he takes us with him on his harrowing journey, which is rendered with exactitude, humor, and lyricism."
—Maggie Nelson, author of *The Art of Cruelty: A Reckoning*

"MacLean fearlessly explores his journey to the edge of madness and his subsequent return to sanity in an unsettling, sometimes riotous, memoir."
—*Publishers Weekly*

"Riveting, sad, and funny . . . Both a sharply written autobiography and an insightful meditation on how much our memories define our identities."
—*Booklist*

"Mesmerizing."
—*Kirkus Reviews*, starred review

"A compelling personal account and a frightful caution to physicians and travelers who continue to place their faith in a very dangerous drug."
—Remington L. Nevin, MD, MPH, mefloquine expert

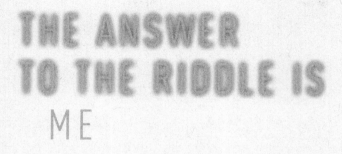

THE ANSWER
TO THE RIDDLE IS
ME

THE ANSWER
TO THE RIDDLE IS
ME

A MEMOIR OF AMNESIA

David Stuart MacLean

Mariner Books
Houghton Mifflin Harcourt
Boston New York

First Mariner Books edition 2015

www.hmhco.com

Library of Congress Cataloging-in-Publication Data
MacLean, David, date.
The answer to the riddle is me : a memoir of amnesia /
David MacLean.
pages cm
ISBN 978-0-547-51927-2 (hardback)
ISBN 978-0-544-22770-5 (pbk.)
1. MacLean, David, date. 2. Amnesiacs — Biography. I. Title.
RC394.A5M34 2013
616.85'2320092 — dc23
[B]   2013026337

Book design by Lisa Diercks
Text set in Monotype Walbaum

Printed in the United States of America
DOC 10 9 8 7 6 5 4 3 2 1

Excerpt from "Ego Trippin' (Part Two)": Words and Music by David Jolicoeur, Vin-
cent Mason, Kelvin Mercer, and Paul E. Huston. © 1993 Warner-Tamerlane Publishing Corp. (BMI), Daisy Age Music (BMI), and Prinse Pawl Musick (BMI). All Rights on
behalf of itself and Daisy Age Music administered by Warner-Tamerlane Publishing
Corp. All Rights Reserved.

  Excerpt from "My Story in a Late Style of Fire" from *Winter Stars,* by Larry Levis,
© 1985. Reprinted by permission of the University of Pittsburgh Press.

Parts of this book have appeared in different forms in *Ploughshares* and on *This American Life.*

For my mom and dad

The answer to the riddle is me and here's the question:

—*De La Soul*

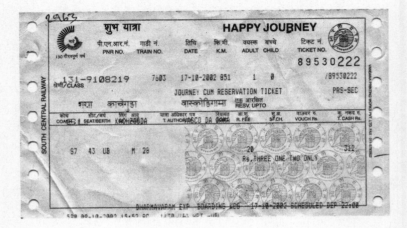

PART ONE

These then are some of my first memories. But of course
as an account of my life they are misleading, because the
things one does not remember are as important; perhaps
they are more important.

—*Virginia Woolf*, Moments of Being

I was standing when I came to. Not lying down. And it wasn't a gradual waking process. It was darkness darkness darkness, then snap. Me. Now awake.

It was hot. My thin shirt clung to my back and shoulders, and my underwear was bunched into a sweaty wad. The heat left the ground in wavy lines, and the air was tinged blue with diesel exhaust. A woman in a burqa pushed past me. A small man in a ragged red vest ducked around me. He was hunched under the massive steel trunk on his back; the corner of the trunk nicked my shoulder as he maneuvered by. I was in the center of a crowd, half surging for the train, half surging for the exits. I stood still. I had no idea who I was. This fact didn't panic me at first. I didn't know enough to panic.

In front of me was a train. A heaving, shuddering train, its engine, half-submerged in smoke, painted a deep red. It blasted its horns, then clanked and panted into motion. People waved to me from open windows as the train shook itself free of the station. I waved back and noticed the whiteness of my arm, covered in hairs the color of straw. I tracked the train's slow-motion progress. As I choked on the bursts of blue exhaust and stared at the receding last car, I wondered if I should have been on that train.

I checked my front pockets for a ticket. Nothing.

Not even a passport.

Now I began to worry. I had lost my passport. I was in a train station in a foreign country without my passport. Then I realized that I couldn't even think of what name would have been on a passport if I had one or what foreign country I was currently in. This is when I panicked.

A man in a small nearby stall clanked a pan against a propane burner. He banged and scraped a spatula against the pan that clanged against the metal burner. The sound was impossibly loud. Louder than the train had been. I wanted to ask the man for help. I didn't want the man to know I needed help. I wanted him to stop banging the pan.

I could feel a heavy absence in my brain, like a static cloud. I couldn't remember anything past waking up. There was a thick mass of nothing up there. My muscles were taut, caught in a constant flinch, waiting for someone, anyone to punch me. I was alone, alone with no idea how far I was from anyone who knew me. I was alone and empty and terrified. I wiped my face with both palms. I blacked out.

I woke up, and I was still standing there on the bustling concrete platform. Not knowing how long I had before I'd black out again, I tried to formulate a plan. There were small monkeys scavenging among the train tracks. Pigeons pecked among the detritus, then flew what they found up to the peaked roof, where they nested in the gaps between the beams and corrugated metal.

A television monitor hung from one of the metal rafters, flickering with information. My neck craned, I watched as unfamiliar letters flashed on the screen. I couldn't read them. Did I forget how to read? I needed it to make sense. If I was going to get out of here, I needed the words to make sense. The screen was old, emitting a low buzz, and the columns frequently twisted from one side to the other, like there was a tug of war among the vacuum tubes inside the black box. The screen went blank, and I was surprised when it came on again that it was filled with something that I could understand. I experienced a moment of exhilaration fueled by the simple recognition of typed English.

The train names, though, were anything but clear. The Janma Bhoomi Express. The Bhubaneshwar Express. I watched the screen as a drowning man watches the arc of a thrown life preserver. I tried to will the words to make sense, to be useful, to pull me out of whatever I was sinking into. But the screen went blank and cycled to an unfamiliar language. Each time it came back to English I experienced the same adrenaline rush. The words continued to twist on the screen. I don't know how long I stared at it. Long enough to draw attention.

Someone tapped me on the shoulder. I reluctantly panned my gaze down from the monitor and saw a young man wearing a peaked cap. He carried a long wooden stick, and perched above his lip he had a slight mustache. The mustache looked unsure of whether it would last till the end of the week.

"Is there something the matter here?" he asked me.

He looked kind. He looked competent. I needed something now that the television wasn't cooperating. Anything resembling comfort or competence would do.

"I have no idea who I am," I said.

Some dam burst inside of me as soon as I said it. I started crying.

The man took a moment to consider his strategy. He finally decided on "There. There." He patted me on my shoulder. "I am a tourist police officer." He pointed to a complicated bureaucratic mandala sewn on his shoulder. "I am here for you. I have seen this many times before. You foreigners come to my country and do your drugs and get confused. It will be all right, my friend."

I was relieved. I should have known. This was the kind of trouble drug addicts ended up in all the time. It was serious, but I was thankful that this police officer had let me know who I was and that I wasn't to be trusted. I knew who I was. He had given me a key to my identity. I didn't have a name, but I now knew the kind of person I was.

"Do you have on your person anything like a passport?"

I shook my big sobbing head, suddenly a puddle again. Prompted by the man's assessment of me, I started to remember doing drugs with an unattractive redhead in a dark apartment. Her ginger face was covered in acne and nickel-sized freckles. Images of her coming toward me twirling little baggies full of toxic stuff flickered in my brain. Cooking. Injecting. Snorting. Scoring. This is what drug addicts do. Then they get lost and end up on train platforms taxing the patience of good men.

"Do you have anything like a wallet on your person?"

I patted down my back pockets, afraid that I would have nothing to report. But out of my right back pocket I produced a brown leather lump stamped with a picture of a cowboy with guns drawn.

"I do," I said. My tears turned joyful. I flipped the wallet open, and there was a New Mexico driver's license. I shoved my forefinger on the square-inch picture. "That's me." I was electric with happiness. I had been found.

"Okay, Mr. David," the man said. "My name is Rajesh. You may call me Josh. You are an American. It will be easier for you to call me that."

I wanted to grab him and dance with him. I had a name and a nationality now. The sterile emptiness of my immediate waking was gone. I bounced from sobbing to smiling in seconds.

Josh pocketed my wallet and grabbed my bicep. "Let us get you somewhere safe."

He escorted me off the platform and into the main hall of the train station, where there was a wall of ticket sellers behind bars who were slowly dispensing with a crush of people who looked like they meant to push themselves through the bars into the ticket sellers' laps. The cavernous room was thick with language I didn't understand. With his hand kindly clamped on my upper arm, Josh pulled me through the hall. Everyone we passed turned and watched.

I was following a man upstairs, the back of his head bobbing as we navigated a dark stairwell that smelled of cement dust. As we spiraled up narrow flights, the landings had rough filigrees of light coming through a pattern carved into the cement wall. His name was Josh, I suddenly remembered.

We walked up an eternity of stairs. On every other landing there was a glass door lit from the inside. Office suites with a slice of air-conditioning coming out from the gaps between the door and the floor. Every floor another business with people bustling inside. Josh kept walking past them all. My legs ached. I was sweating. My underwear chafed me. I was confused, but I knew Josh. I remembered that he was taking care of me. But then I realized that I didn't know Josh. That a man came up to me in a train station and he took my wallet and then he took me to this place, this hot dirty stairwell. I went with him. I was following him. What kind of idiot was I? Part of my brain urged me to run. He was a scam artist. Even if he was real and he was a cop, I was a drug addict. I needed to get the hell out of there. I needed to find the ugly red-haired girl. We got separated, she and I. I was supposed to pick her up at the train station. Or I was supposed to get on a train to meet her somewhere else. I'd botched it because I was dumb enough to lose my mind. Her name popped into my mind.

"Christina," I said. Acne-scarred, redheaded Christina, the perfect partner for squalid drug romps in foreign countries.

I continued to walk behind Josh as my mind spun through all the possibilities. Robbed and killed by Josh, the scam artist. Arrested and jailed by Josh, the policeman. But I kept following him up the stairs. The inertia of confusion overtook me. I trusted his silly attempt at a mustache, an earnest mustache grown by someone not entirely aware of the way other people saw him. The *scritch scritch scritch* of our shoes on the gritty steps echoed all over the dark and narrow stairwell. Where were we going?

We arrived at the top floor. The stairwell opened up to a generous landing. Three bicycles leaned against the wall in a jumble. Bicycles built like tanks. The grit on the ground was the concrete itself, unfinished and flaking off in chunks. Josh yanked the glass door open, and as he did so it screeched against the jamb.

We entered the sudden chill of a highly air-conditioned Internet café. The room was open, with twenty or so computers buried in waist-high carrels. It was empty except for three young men hunched over a single carrel.

From the cluster of men, the heaviest stood up and jogged over to us. He was the clerk of the shop. Josh pointed to me and said something I didn't understand to the clerk. The man looked at me and shook his head. Josh showed the man the card he had taken from my wallet. The clerk took my card, tapped something into the terminal at the front desk, read something off the screen to Josh, and they talked for a moment more. As they spoke, I watched flowers blooming in their mouths and falling down vines toward their feet. The language they spoke was remarkable. The conversation quickly became a thatch of pulsating tendrils. It ended with the clerk waving in the direction of the terminals.

I blinked. The tendrils were gone.

"Would you like some tea?" Josh asked.

I nodded.

Josh whistled at the clerk, who had rejoined his friends at their carrel. Josh ordered the tea as the clerk stared at him blankly. He then punched one of his friends in the arm, a rail-thin boy in a powder-blue button-down. The boy sucked his teeth in disapproval but jogged out of the café.

The kid came back with the tea. Balancing the tray on the carrel's lacquered edge, he passed us each a teacup, sloshing its light brown milky contents onto its saucer. I nearly dropped mine because of the sudden heat.

"Do not try and drink it yet. Let it cool for a moment," Josh instructed me. I took a sip of my tea anyway. It was still hot, but it was the sweetness that scalded me. Sugar and cardamom pods.

I sent an e-mail to my parents containing a message Josh dictated for me:

Mother and Father,

I am in trouble. I am in India and seem to have lost my passport. I am currently very confused and lost. It will be all right as I am with the police, and they are assisting me. Would it be all right if I came home to stay with you? I will endeavor to be a better son and earn your respect back. Please know that I am very sorry that I ever touched these drugs, and this experience has taught me never to do so again. I will be in contact again soon to instruct you how best to assist me in this.

Sincerely,
David

My head hurt. There was too much I didn't know.

The clerk came over, and he and Josh argued loudly. While they jabbered at each other I opened an e-mail from someone named Geeta. There were many e-mails from her dotting my inbox, so I figured she had to be important. Her e-mail read:

David (or should I call you Dah-wid like your watchman?)

I can't wait for you to get here. My landlady is crazy, but she's lent me her scooter for us to use. Do you know how to drive one? If you can, we could go down to the beaches. I have a bikini, but I need a husband around before I wear it down here, otherwise I'm just another Indian American whore. So I'm asking you two things: can you drive a scooter and will you be my husband?

These are obviously very very important pressing questions. So peel yourself away from those scary movies you're always watching (*Evil Dead,* really? You live alone and watch things like that?) and tell me you'll fake marry me and drive me down to a beach.

I need sand between my toes. Stat!

Dinosaur,

Geeta

PS: I've gotten tired of writing OX on my e-mails. I don't know why that yoked mammal is such an affectionate way of ending our correspondence. So I'm substituting something a bit more badass. Prehistorically badass. With teeth!

Her name wasn't Christina. It was Geeta. The woman I was supposed to meet was named Geeta. I hit the reply button.

Geeta,

where

are you?

i am safe thanks to the tourist policeofficer.

where are you?

i"m feeling like i'm ready to go home.

don't have my passport but figuring ways around that.

be good.

I sent the e-mail and closed the browser.

Josh's hand was around my bicep again. He pulled me out of the café and down the steps.

Going down the stairs was much faster than going up. We zipped past the shadows and grit. It was a quick three flights. Down. Pivot. Down. Pivot. Down. We exited into a busy intersection. I blinked away the darkness and found myself in a shock of light and heat and smells. Rickshaws. Cars. Riotously painted trucks belching exhaust. Mopeds, motorcycles, bikes. All tangled up together. All honking in the midday haze. The edges of the world kept peeling up and curling in this heat.

At the center of the snarl of the intersection, inserted in the chaos, three boys popped from vehicle to vehicle, clasping their hands together in routine genuflection, affecting a moment of solemnity, then darting their hands out for rupees. They were identical. Each wore small wire frames with no glasses in them, each wore a short length of cotton wrapped around him like a diaper, each was shaved bald, each had a tiny mustache drawn above his lip, and each was slathered head to toe in silver paint. Silver heads. Silver glasses. Silver dhotis. Silver sandals. Three silver boys dancing in the middle of the street in the middle of the day. Their tiny silver heads glinted as they climbed upstream through traffic like salmon.

Flowers with bloodred blossoms the size of coffee mugs bobbed in a narrow lawn in front of the house Josh and I walked up to. I don't know what vehicle I climbed out of. I was following Josh. I don't know if Josh had his own car or if we had jumped into a rickshaw. Or if I was straddled behind him, helmetless, on a scooter. I blinked out. Pieces of the day kept blinking out, like bad bulbs on a defective strip of Christmas lights. Between the Internet café and the bobbing flowers of this skinny strip of a front yard is a dead lightbulb in a string of memories.

The house was unremarkable other than the flowers, one in a series of concrete slab construction multistory dwellings on a narrow street. It was painted white, and the grit of the city had nestled into the tiny pockets of the cement. The house was built like every other house in the neighborhood, with a raised first floor to make an empty space underneath for the garage. A washing machine was down there, as well as lines of laundry dripping onto the hood of a white hatchback.

Josh knocked at the front door, which was so heavily lacquered that I caught a glimpse of my reflection in the grain. It startled me a bit. Pale, short, and chubby, wavering each time Josh knocked, this stranger was me.

The door opened to reveal a short woman drying her hands on a towel. She looked like she was Chinese, which confused me. I was in India. What was she doing here? I had assumed I was the only one who wasn't like everyone else. This was going to be harder than I thought.

The woman ushered us into a spare living area. The floor was tiled in white marble that also ran halfway up the walls, which were painted white, as was the ceiling. The three pieces of furniture were teak with thin cushions and creaked loudly with even the most minimal movement. I sat on the couch, while Josh and the woman took the armchairs. The room was wide and full of echoes, and each wall had one piece of art rationed to it. The room was arctic and spare, a white cube with anemic sticks of furniture. We could have been inside an igloo, a furnished ice cube.

Within this sparse precision, there was a hip-high slash of color and smell in one corner. Three antennae of incense were stabbed into a green vase, their tips embering into smoke. The table was strewn with all sorts of tiny objects, some shiny, some familiar in shape, some mysterious remnants sitting in small dishes, and a small cardboard box tied with string. At the back of the tiny slice of anarchy were three glossy snapshots. From where I sat I could make out a smile and a thatch of hair in one of them.

"Mrs. Lee, this is Mr. David. He has been having quite the time of it." Josh had put his hat on his knee, and it wobbled as he spoke. "I found him on the platform of the rail station. He is confused and out of sorts."

Mrs. Lee's eyes pierced me. "I am here to help."

"We bring tourists like yourself who have fallen on hard times here." Josh cleared his throat and changed topics so suddenly that I thought I might have briefly passed out. "Would you have something like tea to offer us?"

"Do you want tea? I also have water and a little Sprite."

"Sprite, please," I said.

Mrs. Lee left the room, and Josh leaned toward me conspiratorially and said, "She is a good woman, Mr. David. You may trust her with your trouble."

Mrs. Lee came back with a tray. She handed Josh a brimming cup and saucer, and she gave me a tumbler filled with lukewarm Sprite. I sipped it. It was flat, so the sweetness was heightened, like drinking low-viscosity honey.

Josh blew gently on his tea and took a nibble from the top, then placed the cup down and turned to Mrs. Lee. "Please, tell him your story."

She straightened her posture and began.

"My son was eighteen when he traveled to Singapore. He was going to visit a school friend. He was very popular in school. Handsome. His grades were not the best, but he worked hard."

The room creaked and echoed as she spoke. The room felt empty, as neat rooms do. The emptiness could shake us off if it felt like it. My stomach flipped around inside of me.

"He was only supposed to be gone for seventeen days. Every other day he called. Then nothing. I thought at first that he was being thoughtless. I let a week go by before I began a calling campaign. I spoke to his friend's parents, and they had not seen him. I rang up hospitals. Nothing. I rang the police. Nothing. I went to the airport on the day he was due home. Nothing. I checked with the airline, and his ticket had been refunded. A month went by. I became crazy. Ringing up everywhere. The embassy here. The embassy there. One more month goes by, and a package arrives. In it is a letter from the Singapore government explaining that some bad men had put drugs into my son and that he died. There is also a small bag in the package, containing his ashes. I am a mother with a broken heart. You have no idea what you do to your mother when you put these drugs into your body."

Mrs. Lee then began to cry. I cried right back at her. It was people like me who had killed her son. People like me. I put my hand on hers and told her I was sorry, that I'd do better, that I was done with all of the drugs. Forever. My insides felt like they'd just fallen into an abandoned well. The gray static and fuzz from before was replaced with a black hopelessness. Mrs. Lee took a napkin, folded it three times, wiped at her black eyes, and excused herself.

Josh pulled his cap off his knee and leaned forward, his forearms settling on his thighs. "Now you see why I brought you here. She is a woman who can teach you things." He took a sip from his tea. He smacked his legs with open palms, signaling the end of the lesson. "Is there some way to reach your parents? Do you have a phone number for them maybe?"

I thought for a second. "They definitely have a phone."

"Do you know the number? It would be very helpful if we could contact them. We have sent an e-mail, but it would be best to speak with them directly."

At that moment, Mrs. Lee came back into the room, the napkin still in her hands. "Let me show you where you will be staying."

We climbed a flight of stairs and entered a small room with a bureau, a chair, a mattress on the floor, and a lamp on a small table beside the mattress. An off-balance ceiling fan spastically stirred the air. I walked directly to the window. There was a narrow balcony outside, and I yanked at the glass door to reach it.

"Here. It needs to be unlocked first." Mrs. Lee bent down and flicked a piece of metal, and the door pulled open with a screech. On the street, a man pushed a cart loaded with stacks of paper. He rang a bell as he walked and called out to each of the houses. Four puppies rolled and snapped at one another in the gutter. A woman ironed clothes in a small storefront across the street. There was a braid of wrist-thick black electrical cables coming out of a pipe not five feet from the balcony. They swayed heavily in the breeze and stretched across the street, stitching the buildings together; cinch them tight, and you'd close the open wound of the street.

There was a flutter of movement above me. Three small children chased each other around the open roof of the opposite building. On the building next door to that one, a pair of children stared straight into the sky while fiddling with their hands. I craned my neck to see what they saw. A kite. The string was nearly invisible in their hands and in the sky, but the small patch of color above was clearly leashed to them. Something whirred in my brain. I stared again at the building opposite. On the roof was a small flat.

"That's my apartment," I told Mrs. Lee. Excitement crackled in my throat.

I pictured pushing open the door and finding the squalid flat where Christina and I used to hole ourselves up in while we shot up with heroin and whatever else we could find. The flat was dark; even during the day it was dark. Miserable, with a laughably thin mattress where Christina and I would crash and moan between highs. Standing on Mrs. Lee's balcony, I yearned to go over there, yank the padlock off the door, and enter into my horrible and wasted life. I could see now, though, how bad this was. I could go over there and collect the redhead and get us both some help. Mrs. Lee and Josh would help us. We'd be okay.

"You don't live there," Mrs. Lee corrected.

"I do. I just rented the place."

"No one lives there. It's abandoned."

"Right. It was. Then I moved in." I was sweating with conviction. Why was Mrs. Lee trying to keep me from my apartment? Suddenly, I was awash in paranoia. Mrs. Lee and Josh were the ones who'd drugged me. They were trying to keep me away from Christina, who was also named Geeta sometimes. They were trying to rob me. I wasn't confused. They were making me confused. "That's my apartment," I said again, pleading with Mrs. Lee to let me go.

Mrs. Lee grasped my shoulders and said, "That is not your flat. Do you think you could move into my neighborhood and me not know about it?"

She was right. But if that wasn't my apartment, how in the world was I remembering it so vividly? Now I couldn't be sure of the memories that I did have. Everything was suspect. I was worse than a drug addict—I was nothing. A drug addict could cry over his wasted life. I didn't know what life my tears were for. There was only an absence. I cried for something I didn't know. The braid of black wires swaying in the breeze now asserted itself as a fair ending. I could jump out and grab them. End this. Sizzle away this not knowing, let the people in the street scrape me off their sandals. Send the inky fried residue to my mother in a box. Give it to her to cry over when guests come. Let me jump. Let me end this.

Please.

Mrs. Lee turned me and led me back inside. She handed me a napkin, and I wiped my eyes and blew my nose. She smoothed my hair with her palm.

"You will be fine. Come downstairs. Rajesh has an idea."

Rajesh's idea involved a pocket map of the United States, the telephone, and patience. It took four hours for us to figure out and locate my family's phone number. I don't remember much of this.

What I do remember is the phone ringing. I held the receiver, the plastic warming in my sweaty hand. A man answered.

"Dad?"

"David? How're you doing, tiger?"

This was my father. There was something in that voice, like the smell of unmade bunk beds, of a box taken down from the attic. I was crying again.

"I'm so sorry. I've been a terrible person. An awful son."

"What's going on? Are you all right?"

"Did you get my e-mail?"

"No. I haven't gotten anything from you since last Tuesday."

"I'm in India, Dad. I'm at a guesthouse with the police."

"Is that your father?" Mrs. Lee asked. "Let me speak with him."

"She wants to speak with you," I said. "Can I come home? I want to come home."

"Who? What in the world is going on?" my dad asked. "Of course you can come home."

"I'm so sorry."

Mrs. Lee took the phone from me, and while she talked to my dad, Josh stood up and shook my hand.

"Okay, Mr. David. I must be returning to my post. You are now in capable hands."

I wanted to hug him. I wanted him never to leave. I wanted to sketch a picture of him, to press him between the pages of a dictionary until he was flat and fix him in my diary. "Here is the man who found me," I'd tell people and flip open the book and show them, not a picture, but Josh himself, with that goofy mustache belying any authority he had. I wanted to have him nearby for the rest of my life. Just in case.

I had woken up alone, and this stranger had been kind. What were the chances I'd ever find anyone as kind again? This stranger was the only person I knew in India, the only person I knew in the world. I had done so many things wrong, but this man looked past it all. Because of him, I wasn't hopeless.

I didn't want to be alone again.

Josh shook my hand, slapped his cap on his head, pulled the door shut, and I incurred the first debt of my new life.

Mrs. Lee took me upstairs and put me to bed even though the sun was still shining. She placed a new glass of warm Sprite on the small end table. The mattress was thin and emitted a crunching sound when I sat on it. She wished me a good night and closed the door.

It was at this moment that things went very bad. The room began to twist. It didn't behave. One corner of the ceiling would be pulled down and nearly brushing my lips and the other would be stretched out miles away. Then the corners would swap distances. Floating clouds of color spun and drifted throughout the room. Sometimes one of these clouds would come and sit squarely on my chest, driving all my oxygen out. The little stool in the corner of the room clattered and moved as soon as I stopped watching it. I'd catch it doing pirouettes in my periphery.

Three birds flew onto the balcony and looked at me dismissively. I writhed on the mattress, the ends of the fitted sheets snapping toward me with a *phuff* as they pulled loose. I wet myself.

Things shifted.

I was in a house I knew. I was an old man in my house. The
one I'd lived in for years. The one I'd raised children in. My
house. And it was my birthday. My anniversary. My birthday.
The kids, my friends, my wife, they all thought they were so
damned smart. I'd heard them clamoring around all day. They
couldn't surprise me. I knew they were out there. I placed my
hand on the hollow core door. On the other side of that door
was a wide pink kitchen. Salmon really. A salmon kitchen with
hickory cabinets. An island in the middle overflowing with fruit.
This was all scripted. Be in a family for long enough, and every
move is scripted. I was supposed to walk out of the bedroom
after my nap, go into the kitchen (ignoring all of the snicker-
ing and giggles coming from the lesser hiding places and the
pinwheel of halls and rooms coming off the kitchen), and reach
into the top cabinet for a box of crackers. Once I had the crackers in my hand, I was supposed to say something. A line. Something famously me. Full of wit and an old man's acceptable bile.
I was responsible for saying the line, or at least the first part of
it, and all of my loved ones—my family, friends, wife, dogs,
rabbits, the whole menagerie—would pour out of their hiding
places, and they would scoop me up into their arms and shout
and finish the line for me. I stood there at the door, fresh from
my nap, savoring their anticipation. The little ones, the grand-
kids, oblivious and chatty, aware only of the electricity of this
moment, the phosphorescent infusion of concentrated love into
the atmosphere. My wife—flowing brown hair and almond
eyes—who, once I say my line, will give me the kiss that I get
this one time a year, a kiss full of history and passion. We kiss
every day, but this kiss on my anniversary, on my birthday, on
my anniversary is something special. It's a kiss that says, "You
are here. You are mine. This is ours."

I shifted my palm from the door to the handle and realized

that I couldn't remember what I was supposed to say. I chided myself. An old man's brain. I'd forget my head if it weren't attached. It'll come to me. Just got too excited there for a moment. I can't even remember how it begins. If I knew the first couple of words, the whole thing would rattle off my tongue of its own accord. I just needed to get the beginning.

I figured that what I should do is go out there, get myself to the kitchen, and open the cabinet. I creaked the door open a bit. I peeked out through the slice of dim light coming from one of the nightlights we put in around the countertop, because at this point we seem to be looking for ways to make our electricity bill higher. If I went out there and couldn't remember what to say, how would everyone know to come out? I could've asked them to come out, but that'd be sticking a pin into the party. And she'd know. Christina would know. The kiss wouldn't be what I need. It'd be shaded with her worry. An old man, a doddering fool. Can't remember what he's so damn famous for having said.

I opened the door and peered out, trying to see where people were hiding. Maybe someone would be close enough, and I could get her to give me a hint. Nobody needed to worry. I just needed a hint. It was like a phone number you've known your whole life. Get the first couple of digits, and everything else comes out like a train pulling freight. Someone moved across the hall. I banged the door shut, worried that she might have seen me. My breath caught in my throat. The metal taste of rising panic.

I needed to remember. It was stupid. Something I said all the time. I cracked the door again, but the person was right there. I didn't know her. Some old Asian lady. I banged the door shut in her face.

I fell backward onto the mattress. Old man. Old man. Old man. I accidentally kicked a glass, and it slid across the linoleum and broke against the wall. Stupid old man. Couldn't remember nothing. Stupid.

Someone was knocking at the door. Calling my name. I didn't

recognize her voice. How did she know my name? Was she angry with me for not remembering? For ruining everything by not remembering? The party, the pile of fruit on the island, all the guests, everyone I love, all there for me, and I screwed it up.

There were voices.

The voices were real and coming from a spot just behind my ear, making me twist my head left and right trying to spot whoever was talking. The voices started out giving me hints as to the line I was supposed to say, the one that would trigger all of my loved ones to spring out from their hiding spots and gather me in their arms, cheering. As time went on, though, the voices got nasty. They began mocking me for not knowing the line, for being such an incompetent little turd that I'd gone and forgotten.

"Isn't that just like you," they cackled.

I wanted to find the voices. I flipped the mattress over, knocked over the footstool. The light stayed off, though, and I flopped around in my piss-damp khakis as the terror sizzled through me.

It was at this moment that the door opened and two men I had never seen before walked in and snapped on the overhead light. One was caramel colored and had a TV-news-anchor-worthy pile of silver hair. The other was shorter and darker, with a floppy black Peter Tork bowl cut and a thick mustache.

Stunned at their sudden appearance, I stared up at them from the upended mattress stained with the Sprite and my piss and said, "This isn't going well at all."

The short one began talking. "I am Mr. DeSilva, and this is my friend Sampson. We are friends of your mother. We've been told that you have some bad drugs in your system. We need you to stop taking them."

"I am so sorry about that," I said, wide-eyed and glazed.

Mr. DeSilva helped me up, and Sampson flipped the mattress over and quickly made it, laying a towel over the wet part. The men then helped me lie back down.

"We need you to calm yourself, David. Do not be upset with this turn of events. You are safe. Jesus loves you."

And with that Mr. DeSilva and Sampson knelt beside me and prayed over me.

"Oh Lord, Jesus Christ, our savior, thank you for all of our blessings that you have given unto us. Dear Lord, Jesus Christ, watch over this boy here. Take the devil from this boy, Lord. This boy is a good boy. He needs your help, Jesus. Take the devil from him. Draw the devil from this boy, Lord."

Now that the devil was involved, my hallucinations took a turn for the biblical. Now it wasn't some future version of me and my relationships that were at stake, but my soul, for all of eternity.

"Place your mouth on this boy and suck the devil out of him, Lord."

I was in a conference room with floor-to-ceiling windows, caught in a perpetual sunset. We were up so high that you looked down and all you saw were the tops of clouds. The exterior was an ungodly swash of color.

It was a meeting. There were angels, there were demons, there was an argument concerning the status of my soul and whether or not it was allowable for it to pass into the next level of existence. A man with a mane of lime-green iridescent feathers asked if I was prepared to recite the loyalty oath that would permit me access to the next level of experience.

I stammered. The man with the mane leaned back in his chair and exhaled. This was going to take way longer than anyone had expected.

Sampson leaned forward from his chair and told me that it was all right, that Jesus loved me, that everything was going to be okay. His hair was lovely beyond words, a stack of shimmering silver. This was a room of gods. Just on the other side of the doors were wonders that would blind me if I weren't properly prepared. There was family and friends and the adulation of the universe. I just needed to recite the lines that I was supposed to have memorized.

One of the council members, black with a Hula-Hoop of black hair and gold-rimmed glasses, left, walking right out the door. I was failing. The entire council was disappointed in me. Everyone was tired. This was all taking too damned long. Even Sampson was telling me to calm down. He told me that if I didn't calm down they'd have to move me somewhere else.

I cried and cried.

I snapped awake. Sampson was holding both of my feet down.
His polo shirt had wet patches under the armpits. My arms
were flailing, reaching for the edges of the bed, grabbing at
the coolness of the linoleum floor. I was aware and seizing and
ashamed and out of my mind, all at the same time. Sampson
continued to mutter prayers at me but now through gritted
teeth. Mr. DeSilva was gone. The tiny room was full of a sound
that I realized was my own screaming.

Music started. It was like the voices, perfectly present and absent at the same time. I felt swaddled in the melody but couldn't
find a speaker anywhere. It was a harrumphing collapse of a
melody, all *ooompa doompa* downbeat. It swallowed everything
in the room, making it dance to its rhythm. All other sounds
became a part of the song spinning wild in my head. I recognized it. It was the theme song of a mid-1980s kids' show called
The Great Space Coaster, and its premise was that a magical car
came down from the heavens each day and scooped up three
teenagers and flew them through a rainbow, a black hole, and
a massive fish skeleton to arrive at an asteroid deep in space
where there was a gorilla and a massive pasty rainbow man, and
they taught the kids lessons about sharing and honesty. The lyrics promised a magical ride to the other side where only rain-
bows hide.

"It's the Great Space Coaster. Get on board. On the Great
Space Coaster. We'll explore."

That patch of the song kept revolving again and again
through my brain. I was out of my hallucinations, but this song
sat between me and the world. So when Mr. DeSilva came into
the room, he did so with accompaniment. It was that music that
kept me from hearing what he said to me when he leaned down
close and squeezed my shoulder. It was that music that the two
men in green scrubs danced into the room to. And as they lifted

me onto the gurney and strapped me down, it was all choreo-graphed to "the greatspacecoastergetonboardonthegreatspace coasterwe'llexplore."

I was hoisted down the flight of stairs, through Mrs. Lee's ice cube living room, and into a waiting ambulance outside. The coffee mug blossoms of her otherwise anonymous house bobbed in the wind as I was spirited away.

"THE HOSPITAL DOES NOT BEAR ANY RESPONSIBILITY FOR REPORTS WHICH REMAIN UNCLAIMED OVER 30 DAYS"

APOLLO HOSPITALS ENTERPRISE LIMITED   QRF-BL-0901-01
JUBILEE HILLS, HYDERABAD - 500 033.
PHONES : 040-3607777 FAX : 040-3608050 **"ASK FOR HEALTH INSURANCE BROCHURE TODAY"**

OUT PATIENT COUNTER

OTHER PROCEDURES

No.
Name : CS240819   Date :
MR DAVID MICLINE   Sex: Male Age: 28 Years   Time : 18/10/200 Hours:
                                                            3:43:01 PM
Doctor : AMBULANCE CHARGES   600.00

With the theme song for *The Great Space Coaster* still rattling about in my head, I was wheeled into a green room and besieged by a pack of women. Their hands were all over me, pushing my legs and arms down. One woman had a needle, and she kept sticking me with it. I flinched every time so she couldn't dose me with whatever poison she had. Why couldn't they just leave me alone? She kept stabbing me again and again. I forced myself up and swung a wild punch. It connected with the jaw of a thin nurse. She fell down. The tremor of the punch traveled up my arm and sat in my chest like an anchor. I screamed, "No one touches me!" Finally they found the four-point restraints and cinched them down. Hard. A burly nurse came up and, dispensing with all bedside manner, pinned me down and jammed a syringe into a vein.

This drug immediately severed any connection between my brain and my body, and my brain set sail on its worst voyage yet. It was as if someone had scraped all the toxic mess off the inside of my soul, all my personal asbestos and idiosyncratic carcinogens — my doubts, my fears, my failures and insecurities — everything that had caked up inside my neurology, and then force-fed it all back to me until I gagged.

God showed up.

He took my bicep in his hand, and I was thrust wholly into the cosmos. We flew through space together, God and I. But it wasn't giant stars and asteroid belts; it was mostly blackness. Making wide parabolas of exploration, God showed me all of the nothingness of his creation. I couldn't feel movement, but nothing really changes in space. Pricks of light maybe, but very far away. There was a wide hollowness. I hovered in a vast blank nothingness. I was simultaneously scared and at peace. The quiet was deafening.

God showed me Earth, a dumb globe hanging in all the blackness. He then showed it to me in four dimensions. It was unbelievably beautiful. The fourth dimension, God told me, is Love. I saw the world and felt the joy of loving it all, atoms to Alps. Everything was suffused with twinkling idiosyncrasy. It was pure ecstasy to see like this. God told me that it was simple; I could stay with him and see like this for the rest of eternity if I just told him the quatrain he asked me to remember before I was born.

God waited for an answer.

I didn't have one.

He was disappointed in me.

He reached over and flipped on the light. We were in a wide hangar. Corrugated walls, vaulted ceiling, concrete floor—cold on my bare feet. I looked at God, sitting in a director's chair. He was Jim Henson, and that made so much sense to me that my eyes bulged with the truth. Of course God would be Jim Henson. He was fatter than I remembered, but the beard, the eyes —God's gorgeous playful eyes—it was Jim Henson in a nightgown.

"I make it all from here," he said.

One of the wide hangar doors rolled open, and I could see asphalt and palm trees.

"Hollywood?"

"No," Jim Henson said, making a sour face. "Burbank."

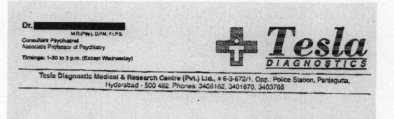

Dr. ████████████
M.D.(Psy.), D.P.M., F.I.P.S.
Consultant Psychiatrist
Associate Professor of Psychiatry
Timings: 1-30 to 3 p.m. (Except Wednesday)

Tesla DIAGNOSTICS

Tesla Diagnostic Medical & Research Centre (Pvt.) Ltd., # 6-3-672/1, Opp.: Police Station, Panjagutta,
Hyderabad - 500 482. Phones: 3406162, 3401670, 3403765

Hyderabad,
18.10.2002

TO WHOMSOEVER IT MAY CONCERN

This is to certify that I have examined Mr. David MacLean, male aged 29 years, citizen of America, at 2.30 p.m on 18.10.2002 in the Emergency ward of Apollo Hospital, Jubilee Hills, Hyderabad. I have found him to be suffering from Acute pleomorphic psychosis and am of the opinion that he is not in a position to give valid informed consent due to his mental condition.

Reg.No: 18413

I was lucid. Something snapped, and I was lucid and in a green room with a yellowing white cotton privacy screen. A man in a white coat sat at the edge of my bed. He was writing something on a chart, but his pen wasn't cooperating. He scribbled it in the margins trying to loosen the ink.

I knew who I was.

I felt washed with it.

I knew that this wasn't normal, but I didn't panic. I knew that I'd be okay. It was like being slipped back into a warm overcoat, smelling your own smell, absorbing your own radiated heat. I was calm. I was tied down, cloth and leather straps on wrists and ankles, but I was calm. My tongue was dry and big in my mouth. On the small wooden table next to me was a small plastic pitcher and cup, both red.

I said to the doctor, "Thank God I took all that acid in college, or I'd really be screwed."

He looked at me with professional kindness.

Something snapped. A bit of tinsel crinkle and a shimmer, and I was gone again. Swimming in infinity.

Strapped down, I wrestled with the riddle God had given me, the quatrain he'd asked me to recite, the one that was the secret passcode into the fourth dimension. I had the first bit. I knew the first line. God had tipped his hand on that one. I was on my way to my family/place among the angels/eternal consciousness, but something was stopping me from completing it.

"If you can't understand that the universe was created by Jim Henson in a studio in Burbank . . ." It was an "if, then" construction. I'd figured out half of it. I said it again. Then again. It sounded right. I was giddy with my progress. Not long before, my soul had been dangling over the precipice. Now I knew half of the riddle of the universe. I just needed the second part.

There were shapes. Blurry shapes hung over me. I was no longer in the hollowed-out universe. I was no longer in a cosmic aerie. I was in a bed, thrashing out a problem, and the shapes were cheering me on. One was dark with glinting eyes. He called me "Hero." He had an Indian accent.

I was in India. India was part of the answer. The most Indian thing I could think of wasn't Gandhi or blue gods or the Ganges. It was masala dosa, the wide, thin rolled-up pancakes of southern India.

"Fuck masala dosa," I said to myself and laughed. God is Jim Henson. Jim Henson was funny. The riddle was gonna be funny. "Fuck masala dosa," I said again and laughed. Then I said it again.

"If you can't understand that the universe was created by Jim Henson in a studio in Burbank, then fuck masala dosa, you've got . . ."

I was two-thirds of the way there.

Fuck masala dosa.

After two days of hallucinating, the olanzapine and lorazepam had finally mellowed me enough to be unstrapped. The nurses brought me things to keep me occupied: newspapers, pens. Convinced that I was still failing a cosmic soul pop quiz, I crouched over the newspapers, believing that the answer was hidden somewhere in them. I was failing a test of my soul, but I was no longer punching people because of it. I circled words and drew complicated diagrams of sentences I found encoded in the disparate articles. The flat newsprint was overlaid with my fevered brain's belief that there was more being communicated. And what was being communicated was essential to me for my soul's progress to eternity. I just needed to work harder to figure it out.

I scratched and scribbled, making connections, trying to conjure spiritual luminosity out of the Hyderabad newspaper, the *Deccan Chronicle*. My biggest problem was that my pen would punch through the newsprint, dragging a rip into the page. When this happened, I'd start crying.

I shared a room with another person. I'd see his shadow cast onto the privacy screen between us at night sometimes, but I never saw him. His son came over to see me all the time. His name was Amol. He was ten and eager to practice his English.

The room, about ten by twenty feet, was painted the same color green as the scrubs the nurses wore, just dustier. The ceiling was black, with spiderwebs in the corners. These spiderwebs moved when I wasn't looking. I'd catch them creeping closer to me out of my peripheral vision. Sometimes the spiderwebs disappeared entirely and the corners were clean and clear. They were sneaky that way. The floor was tiled, but seemingly tiled by five workmen working from different plans. There was no pattern to the floor. Colors, shapes, sizes, all shoved together. When I wasn't staring at the newspapers, I was staring at these tiles. The answer could be anywhere. The world was a sentence that I needed to read in order to graduate to the next level of experience. But nothing was making sense.

I shouted a lot.

I kept waiting for the door to open and for the police to appear. I was waiting to be held accountable for all the things I couldn't remember. I had done such horrible things to people, hurt them, made them cry, abused all of their hard work with my violent disregard. If only I could put together the answer to the riddle Jim Henson had asked me, then I'd be free. I'd keep all the evil away, even the awful stuff that I had done, if I could only finish the riddle.

My room became a revolving door of visitors. Patients from other wards came to visit me. I felt so guilty that I refused no one. Anyone who asked was given the cigarettes that kept showing up on the tiny table next to my bed. Strange men in hospital gowns sat on my bed and practiced their conversational English. I was going to rebuild my life by being good. Nurses lingered at my bedside, gave me heaping plates full of curd rice, and teased me by asking if I wanted some masala dosa. One nurse told me that I was the most entertaining psychotic that they'd ever had.

I'd black out and then snap back awake in no discernible pattern. I'd wake up walking, wake up in the middle of a conversation—the other person looking at me expecting me to finish a sentence I didn't remember starting.

There were moments when everything was clear. I was in a mental institution. In India. This was weird, but I was safe. Everything was going to be all right. I had doctors who squeezed my shoulders and called me Mr. David. There was progress. The worst was over. Then there'd be a sparkle in my periphery, and I'd black out.

I wasn't sure where I'd woken up this time, but there was water in a plastic cup on a table next to my bed and a small brown boy peeking around a privacy curtain at me.

The boy was biting the curtain, and he was rocking back and forth.

"Amol," a man's voice behind the curtain said. "Don't bother the American."

I had an IV coming out of my arm, and I was wearing a hospital gown.

I woke up, and the sun had shifted on the wall and the plastic
cup was gone. A man who looked like Jim Henson, but fatter, was sitting in the chair beside me. He was a white man, midfifties, in a kurta soiled around the armpits, and he was smoking.

"At least you're not tied down anymore," he said.

40   I woke up again, but maybe that's wrong. I never remembered
going to sleep, so maybe I wasn't waking up during all of this.
Maybe it was just flashes of lucidity. My blood felt heavy with
all of the medications coursing through me. When I woke up
in the middle of conversations, I'd apologize to whomever I was
talking to. I was always apologizing. I'd done so many terrible
things.

I spent five minutes moving my jaw. The man behind the privacy curtain was snoring. I pushed my tongue around my teeth. They tasted funny. The plastic cup was back. Directly across from me was a door with a glass window embedded with wire hatching. I could see people through the window, moving back and forth. Women looked in and stared at me until I noticed them, and then they disappeared.

A doctor came in. He saw that I was watching the window.

"Don't mind the nurses," he said. "You gave them a fright."

"What am I doing here?" I asked.

"Recuperating," he answered.

"My teeth taste like paste."

"Can you tell me about these things that you have written?" He knelt by a pile of newspapers and picked one up. For a moment, I thought I was a journalist. On the papers, between and over the printed words, were scribbles and diagrams. They covered five pages of the paper. Blue pen and black pen. The newspapers were soft from being overhandled.

"I did this?"

"You'll be fine. I was just curious." He dropped the paper onto the chair. "You are under the care of Woodlands Neuropsychiatric Centre."

"I am in India."

"You have had quite a time here, it seems."

I woke up, and Jim Henson was back. He hovered over me, putting the back of his hand against my forehead.

"Do you need anything, Dave? Like soda or candy or magazines or books or cigarettes?"

It was too long of a list from a man I didn't know. I said, "Cigarettes."

"Cool, man. I'll be back. You just get better. Take it easy. I don't know what they're telling you, but you're going to be fine. Get that poison out of your system. That's important." He grabbed the IV bag and tried to read the label. "I don't know what poison they're pumping into you now, but that other stuff will be gone soon." Jim Henson reached into a bag slung around his neck. "I brought you some of my poems. The ones I was telling you about. The cycle? You know, maybe you could look at them."

He slid the sheath of papers under my arm.

I woke up crying. The room was dark. I lit a cigarette and
watched the little red dot float toward my mouth, glow brighter,
then float away. When I turned my head it felt like my brain
took a few seconds to catch up. I crushed the butt out on the bed
frame, little flecks of red falling to the floor, and lit another one.
I smoked until dawn.

I often woke up in the middle of conversations with Amol. I could hear Amol's father's voice, and sometimes, when he had his lamp on at night, I could make out his silhouette through the privacy screen, but I never saw the man. It was a room with four beds, but two were empty. I had the one by the window.

"We're going to America. Right, Mr. David? You're taking me to America?"

"Sure. I guess." He seemed like a good kid, and America was such a big place. How hard could it be to get to America?

"Amol, I need you. Please come here."

His father spoke only English in the room.

Jim Henson — I was too embarrassed to ask his real name
since he seemed to know me so well — was sitting by my bed
and chain-smoking with me. His dirty kurta was stretched tight
over his stomach.

"Dave, I went through this, man. You get to this point where
you just feel like everything goes awry, and it's like you realize
that the rest of the world is crazy, and that's when they lock you
up. It's like they have this sensor that gets set off when you realize how full of it everything is, and they come and get you and
lock you up until you start thinking they're important again."

Three birds flew up and sat on the ledge. Their wings, brushing against the windows, made clicking noises.

Jim Henson tapped his cigarette out and then lit another.
"But guys like us know better. Guys like us know how to fake
it. Let 'em think we believe their nonsense. Let us take what
we learn here, and take it out there." He held his hands like he
was holding a brick and moved them from left to right. "That's
when and how the world gets changed. It's the only way it's ever
been changed."

Jim Henson examined the tip of his newly lit cigarette.
"Guys like us."

It was night. Time was a roulette wheel. Any moment could be any moment. Close my eyes. Instant time travel. Right now it was pitch black. There was a fluorescent light from a building next door, casting everything in that grainy green light. As I smoked, I noticed bruises on each of my wrists, purple stripes three inches wide. I slid off the bed and moved as close to the window as my IV would allow, and marveled over these bruises. I wanted to stick a lit cigarette into them.

Amol stood by the window, naming things.

"There is an oxcart. There is a vendor. There is a car. There is a girl. There is a man." He turned and smiled at me. It was nice to have a friend.

His father called out from his bed. "That is fine, Amol. But what are these people doing?"

"The girl is holding hands with a woman. The man is drinking tea. The ox is eating the refuse."

"Trash, Amol," his father said. "Americans say trash."

Jim Henson came back often. There were always at least three packs of cigarettes on the table and fresh piles of poetry. Amol started running cigarettes to his father.

"Ask Mr. David if I could have two cigarettes."

I would have one of the cigarettes lit by the time Amol cleared the curtain. Matches weren't allowed on his side. We'd gotten in trouble before. In our room, I was the one with match privileges.

Amol's father had a visitor. They spoke briefly. I heard the
crinkle of newspaper more than their conversation. The man
looked over at me as he left. He was well dressed. Gold spec-
tacles worn low on his nose. Starched shirt tucked tightly into
his pressed slacks. He came over and shook my hand, beaming.

"And how are you finding our country here, sir?"

"It's great."

"Have you seen the Taj Mahal yet?"

"I'm not sure if I have."

"It's wonderful. Go during a full moon. I think it's more expensive for foreigners, but surely worth it. Truly spectacular."

Did he not notice my hospital gown, did he not see the bruises
on my wrists from the straps, the IV dangling from my bedside
that was topping off my brain with chemical solution?

"Are you a Christian, sir?"

"Sure, I guess." I felt a twinge at this mention of religion but
didn't know what to attribute it to.

The man jingled change in his pocket as he spoke. "It is during these difficult times that our Lord shows himself in his true
glory." He leaned in conspiratorially but didn't lower his voice.
"Be glad you're Christian, son. These damned Hindus were
born to suffer."

I didn't dream at night. The olanzapine made sure that I experienced nothing in my brain that didn't come through my eyes and ears and nose. It was hard to think, much less dream. Consciousness was on or off. When it was on, I was groggy and confused. When it was off, I drifted in a place as unremarkable and viscous as wood glue.

At any moment, they were going to haul in Christina. I had figured that they'd thought that my mind was too brittle to handle
the awful things I had done. When I finally got better, they'd
reveal my catalog of crimes and convict me. All the drugs I'd
done, the nurse I'd punched, the apartment I'd forgotten to keep
up. I had found a way to be worried about getting better. My
present state was perpetual anxiety, and the idea of progressing
past it promised more anxiety. At the end of the tunnel would
be a different kind of darkness. Christina'd be brought in, arms
linked together behind her, and everything would be revealed
to me. I just had to get well enough to handle my guilt.

"I see a man in a dirty red shirt peeling a green banana and then proceeding to eat the green banana in a quick manner."

Amol's English had progressed. It'd be metaphor and simile next.

I was eating. The nurses laughed about curd rice and how much I loved it. I got extra portions of it, along with mango pickle. I shoveled it in. I wasn't sure if everyone used their hands in India or if they didn't want me to have a knife and fork.

"Amol," I said around my fingers. "Soon you're gonna tell me what they're thinking."

He came and sat on the edge of my bed, swinging his legs, bouncing them off the bed frame.

"What's it like not to have your memory? Is it fun?"

"It's not fun." I put the plate on the side table. "It's awful."

"You don't know anything at all about yourself? You could be a murderer and not know it."

"Do I look like a murderer?"

"Or you could be a famous man who came to India to give money to young boys so that they could complete their studies at the best schools."

"If only I could find a deserving young boy."

"Or you're a cricket superstar. The fastest bowler ever seen."

"I don't think they play a lot of cricket in America." I pulled a cigarette out and tapped it against the table.

"Have you always smoked?"

I shrugged and lit it.

"Amol," his father called from behind the curtain. "Please ask Mr. David for two cigarettes."

"Shouldn't I be talking to someone?" I asked the doctor.

"You want to talk?"

"Not really, but shouldn't I be talking about this to someone?"

"We're not that kind of hospital." After making a few ticks with his pen, he closed a folder on his clipboard. "You are on the appropriate medications."

"Shouldn't I be talking to someone about something?" I lit a cigarette and then noticed that I already had one going. "You know, about the stuff I've done?"

"You've done nothing. It's not that kind of problem. You had an allergic reaction to a medication you were taking. An unpleasant side effect."

"Allergic reaction? I thought I was on drugs."

"You were not taking drugs; instead you were taking a drug. Nothing illegal."

"Am I crazy?"

"Not any longer." The doctor squeezed my shoulder. He was a big one for that gesture. "You will be fine. Your memory will return, and all of this will fade."

Panic rose in my throat. "I'm going to forget all of this?"

"This confusion is what will fade." He closed his notebook. "We can stop having this conversation every few hours."

A woman was in my room. She was white, in her early fifties, had a sensible helmet of perfect curls, and wore an elaborately beaded kurta. She and Jim Henson were talking amiably. Their conversation was intimate. They were friends. She noticed that I was watching them.

"David, I wanted you to know that I placed a call to the embassy in Chennai. Your parents are on their way. Fulbright representatives will meet them in Mumbai, and then they'll catch a connecting flight here. They should be here by tomorrow." She spoke loudly, as if I were a little deaf.

"Where are they staying?" Jim Henson asked.

"The Taj," the woman said.

"Nice."

The two of them exchanged a look that wordlessly expressed their opinion about the kind of people who stayed at the Taj.

"The phone call was eighty-seven rupees. I took the money for it out of your wallet, all right? You only had a hundred, so I'll owe you, okay?"

I nodded my head. I reached to check my pocket, but I was in a hospital gown.

A nurse brought me a plate of curd rice. The plate was metal and had sections for several different portions, but each section just held curd rice.

"I brought extra for you."

On the other side of the privacy curtain, I could hear another nurse scraping a utensil against a metal plate and urging my roommate to eat. He'd been tied down again.

I dug into the pile of milky sweet rice. The taste of it counteracted the bitter weight of the medications.

"I put some pickle there for you. It is a little spicy only."

I mixed the orange hunk of okra into my next bite. It was sour but cut against the sweetness. I beamed at the nurse. She had made my favorite meal even better than before.

She straightened my pillow and took my empty plate from me, holding it against her hip. She smiled. "Maybe soon, Mr. David, you will want some masala dosa, maybe?"

The nurse on the other side of the curtain started laughing.

I woke up to a cigarette burning between my fingers. I woke up to Amol's fingers touching my hair. I woke up to the sound of monkeys fighting on the tiny ledge outside my window. I woke up to an incredibly dark-skinned man calling me "Hero." I woke up to Jim Henson advising me against electroshock. I woke up hunched over newsprint, forgetting the obvious connection that had been there just a second before. I woke up to a tiny Indian woman in glasses telling me that my mother loved me very much. I woke up to a Spanish woman telling me that Chekhov was her very first love. I woke up to Mr. DeSilva's cool hand on my head and him introducing me to his wife, who stood behind him and watched me with her arms crossed. I woke up and asked strangers about Christina. I was worried about her. No one had any information on her at all. They acted like she didn't exist. I faded in and out of my life like I was channel surfing, but there were only two channels: the wood-glue nothingness that I kept slipping into and the continuing mysteries of who I could possibly be.

PART TWO

"How can I tell," said the man, "that the past isn't a fiction designed to account for the discrepancy between my immediate physical sensations and my state of mind?"

—*Douglas Adams,*
The Restaurant at the End of the Universe

Dr. ███████
M.D.,D.P.M.,F.I.P.S
Regd. No. 10655.
CONSULTANT PSYCHIATRIST
Honorary Consultant to :
Heritage Hospital, Vijaya Health Care,
Mediciti Hospital, S.C.Rly HQ Hospital,
Woodlands Asha Neuropsychiatry Centre,
Sai Kidney Centre.

Mr David Maclean   20¹⁵ Oct 02

28 yrs M.

Advise.

① Tab OLEANZ 5 mg   1 Tab morning
   (OLANZEPINE)   1 Tab night

② Tab ATIVAN 1 mg   1 Tab twice a   NIGHT
   (LORAZEPAM)   day
                        (one more at night if
                          Sleepless / Restless)

Review on ~~22/10~~
         22ⁿ Oct. '02
         by 10.00 AM

8.30 AM to 10 AM:	10.30 AM to 2 PM:	2.30 PM to 4 PM:	6.30PM to 9PM (Mon to Fri)
Woodlands Hospital,	Sri Sai Kidney Centre.	Heritage Hospital,	Vijaya Health Care
Barkatpura	Seeshmahal Road,	Civil Supplies Bhavan Lane.	Beside PrasanthTheatre, Clock
Ph : 7560918,	Ameerpet. Ph : 3756789,	Somajiguda	Tower, Secunderabad.
6578764	3756644, 6586996	Ph : 3379999, 3379201	Ph : 7701344

For PSYCHIATRIC EMERGENCIES please contact WOODLANDS ASHA HOSPITAL, BARKATPURA
Phone : 756 0918, 657 8764, Mobile : 98480-22861

The birds returned. It was morning. They were pigeons, and the sun made their iridescent neck feathers flicker from blue to pink to purple as they fought over a foil bag. I'd had another big serving of curd rice for breakfast. The nurse stayed to watch me eat it and complimented me on how well I managed eating with my hands. She wiped my mouth and cleared my plate. I had a brief moment of sadness when she left the room. People were always leaving.

I looked at my IV bag and wondered if it was putting fluids into my body or if it was taking them out. Was the clear bag attached by tube to my wrist pulling the badness from me and, if so, was it consequently full of my insanity?

I wasn't sure how long I'd been in the hospital. Time stretched and condensed in ways that I couldn't measure or predict. I could have been there overnight, or I could have just as easily never been anywhere other than that room in my entire existence. I reached over for my cigarettes on the small side table, and my parents walked in.

Some motor in my brain spun and sparked a blue arc of electricity between two exiled neurons, and *pow:* recognition. It was like a day spent sorting through your attic compressed into one millisecond. They were my parents. They looked like hell.

We made a little tent over the bed with our bodies and wept. I smelled the humid travel smell of them. I was instantly nostalgic for the moment when they walked in. There was a light in that moment, a shaft of pure promise. We let go of the hug, and I was confused again. I wasn't sure if we'd done this a hundred times before. Were they always visiting me in places like this?

My mom said she was sorry. I said that I was sorry. My dad pushed his glasses up in what I instantly knew was a famous way of his when he squinted back tears. It was wonderful to recognize the particular way my father cried.

Dr. Chandra was with them.

My mom sat on the edge of my bed and smoothed my hair as the doctor talked quietly with my dad. She pushed her thumb into the space between my eyebrows, and I recognized that gesture, too. It was something she'd done my whole life, wordlessly telling me not to worry so much. I still didn't have my memory, but I now had an outline of myself, like a tin form waiting for batter.

My eyes hurt. My parents being in the room with me was doing things to my brain. I was groggy with all of the medications. It was as if my emotions were smothered in layers of saran wrap, but I could feel the rightness of them: these people fit in my life, and I fit in theirs. They were a big piece to a puzzle. My mom hugged me again, careful of the IV.

"We're going to get you home," she said, touching my forehead, my cheeks, the sides of my neck.

The white woman in the beaded kurta came in almost immediately. She acted like she'd just been stopping by and was surprised at her good luck at finding Mom and Dad there.

She shook both of their hands and introduced herself as Dr. Pat. She had a wiry thinness about her, and her veins stood out on the backs of her hands, thick as pencils, almost as if they were trying to jump clear of her body. She hugged my mom and used plural pronouns about my recovery: "We're all doing much better now." She talked with my dad and the doctor in the corner while my mom reached down into a bag and pulled out a shoebox. For some reason, I resented Dr. Pat's presence. I didn't need her anymore. My parents were here.

"Betsy sent these along," my mom said.

Inside the box were two dozen cookies wrapped in wax paper. I munched one. Cranberries, chocolate, and walnuts, a universe of taste away from curd rice. At the bottom of the box, insufficiently shielded by the wax paper, was a manila envelope spotted with cookie grease. I shook out the contents of the envelope.

There were dozens of photographs, all of them with me somewhere in them. If each photograph represents one sixty-fourth of a second, I held maybe two seconds of my life there in my hands in that hospital bed. Not much in the scheme of things, but at that moment it was everything.

There was a picture of me, my blond hair sticking up in a thousand directions. I had an empty industrial-sized black garbage can in each hand, and my mouth was wide open in a howl. There was a picture of me, legs crossed, in a suit jacket and a kilt. There was a black-and-white picture of me caught in the middle of a pirouette on the hood of a Toyota station wagon. There was a picture of me hugging a black-haired woman with an amazing nose and a gorgeous gap between her teeth. There was a picture of me in a tux with a Frisbee tucked under my arm.

I held up a picture and asked my mom what I was doing with my face like that: my eyes bugged out, my mouth screwed up.

"That's just something you do, David."

Amol came over from his dad's side of the room and introduced himself to my mother and then to my father. He shook their hands and said to each that it was very nice to have made their acquaintance.

"I'm helping him with his English," I said to Mom. "He wants to come to America someday and visit us."

Amol turned to my mother. "David tells me that you are working with education. I am very interested in education. David has also said that you might be able to help me with coming to America?"

My mom's kind face fell apart for a millisecond as she tried to figure out what promises I had made to this kid. Recovering a bit, she chatted amiably about the process a person would go through in order to study in America as a university student, but she added that she thought it unwise for a young person to be away from his own family during his formative years.

"I only have my father left," Amol gestured to the privacy curtain, "and he is a suicide, which is why we are here only."

I tapped out a cigarette from the pack on the little table and lit it, sucking in the smoke like a pro.

My mom immediately stopped talking with Amol. "What in the world are you doing?"

I told her that it was no big deal. That this guy was always bringing them by for me, so it wasn't like I was spending any money or anything.

She explained that I'd been diagnosed with asthma while I was in third grade and had struggled with it my entire life. "We used to send you to asthma camp in the summers," she said, exasperated. She checked herself. "We'll talk about it later."

Jim Henson walked in, along with a very dark man with black hair parted down the middle and spread out like wings. The room was becoming uncomfortable. Jim Henson introduced himself to my parents. Apparently, he had a name, and it was Richard, and the wing-haired man introduced himself as Veda.

Veda came over to me immediately after shaking my father's hand. "How goes it, Hero?"

"My parents are here," I told him. "They're going to take me home."

"My students will be very upset." Veda turned to my mother, clasped his hands behind his back, leaned toward her, and said, "Many of my female students are quite taken with David."

Here was an absence, a complete erasure. I had no idea what he was talking about. I was a teacher? Girls liked me? Each new example of emptiness was like a glass bottle breaking inside me.

On the other side of the room, Richard, Dr. Chandra, and Dr. Pat had cornered my father. Richard had a sheaf of paper stuffed poorly into a folder, and his kurta pocket bulged with two packs of cigarettes. He was in his midfifties, carried a sour smell with him, and talked about his poetry as if it were particle physics. I could see my father squirming away from the man. I recognized that, too: my father not wanting to be in a conversation.

It was a Monday afternoon, and as the call to worship rang out from the tinny speakers of the three mosques surrounding us, the world had come to buzz with concern by my bedside in the asylum. The room hummed with conversation, and all the while each visitor kept one eye on me as I flipped through the pictures: me, the glazed subtext to every conversation. Amol squirreled up next to me on the bed, careful of my IV.

"Who is this?" He pushed his finger on the face of the girl with the great nose. It was a large nose, a nose that dominated her face, but one that lifted all of her features to a pinnacle of beauty. How boring she would have looked without that nose. Just another blandly beautiful woman. That nose, that gap in her teeth marked her as precious. We were in a parking lot. She carried an SLR camera with a huge lens, sunglasses perched on her head, wearing an oversized black sweater. My hands were jammed deep in my pockets in a cool boy pose.

"I don't know."

I flipped to another picture. In this one I had my arms around a woman with piercingly blue eyes. She had a pixie haircut and a purple cardigan, and I was acting like I was about to bite her neck.

"Who is she?"

"I don't know," I repeated.

He asked again and again, each time placing his finger on a face adjacent to mine. Smiling faces. Beautiful faces. Maybe girlfriends, maybe cousins, maybe college roommates.

I don't know.

I don't know.

I don't know.

It was like he was lancing abscesses in my memory. Each face he poked with his finger caused a pop in my head. Where a memory might then crystallize, there was nothing.

I held up the pictures, interrupting Veda and my mom. "Which one of these is Christina?"

Mom's face was blank.

"Christina? Or Christine?" I asked.

"I've never heard you talk about someone with that name," my mom answered.

"I think I love her," I said while flipping through the pictures. "Maybe her name is Geeta." I didn't know if I should tell my mom about how Christina and I stayed in terrible apartments and shot drugs right into the veins in our arms. I wasn't sure how much my mom knew. How much I should spare her.

My mom reached and rubbed my cheek between her thumb and palm. "I know Anne and Sally are both very excited to see you."

I nodded and said that I was excited to see them as well.

Whoever the hell they were.

My parents came the next day and checked me out of Wood-lands Asha Neuropsychiatric Centre. We settled up the item-ized bill first. Three days of hospital care: medicines, admission charges, registration fee, bed/room charges, injections, IV flu-ids, disposables, investigations, treatment charges (counseling charges, nursing charges, consultation fees), food and other charges, and miscellaneous. It all added up to 1,900 rupees even, or about US $40.

I was given a pillowcase full of the possessions I'd had with me. My parents had swung by my flat and picked up a change of clothes and my digital camera. As we left the asylum, I started taking pictures of everything: the helter-skelter patternless tiled floor; the cream-colored spiral walkway around the court-yard with its broken fountain; the patients in the courtyard, di-vided—the insane men on one side, insane women on the other —the gates; the rickshaw we climbed into. I soaked the camera in images. If I lost everything again, I'd be up-to-date.

The three of us squeezed into the tiny black and yellow rick-shaw, me in between the two of them. My father was clenched tight. Mom, on the other hand, was overly chatty. As we circled the ring road that surrounded the town lake, dodging bicyclists, cars, pedestrians, trucks, mopeds, motorcycles—the whole gamut of weekday traffic—she told me that I wasn't a drug ad-dict, that I hadn't been out of contact with them for years, that I really didn't need to keep apologizing.

She told me that I was in India because of a fellowship from the US government, that the doctors she'd talked to back in Ohio thought that what had happened was likely the result of an antimalarial drug I was taking. My brain took in this infor-mation like a cruise ship effecting a U-turn. I wasn't to blame? It was impossible. My mom must have been lying to me, doing what the doctors were doing, protecting me from the truth.

"What did I get a fellowship for?" I asked.

"You're a writer."

It was news to me, but it now made sense that Richard had always been at my bedside wanting to talk about poetry. "I must be pretty good," I said, "if the government paid me to come over here."

My mom grabbed my knee. "Don't be too flattered. This is the same government that pays $600 for a toilet seat and pays farmers to not grow crops."

And we laughed. She and I laughed together. At me. I recognized it. The banter had a kind of music to it. We kept at it, making jokes as the rickshaw flew into the oncoming lane of traffic and Dad nearly snapped the handle off his side.

There's a giant statue of Buddha in the center of the lake. He stands ten meters tall, with his palm outstretched in a gesture of peace. When the statue was first being installed, it had capsized the boat that was carrying it and killed three men before sinking to the bottom of the lake. During the recovery effort for it two years later, another two men were killed. I'd learn all of this history later. At that moment, with my father about to grind his teeth down to the nerves and my mother and I laughing away, I saw the stone Buddha with his palm stretched out. He was giant and heavy and real and seemed the epitome of peace. While our rickshaw ricocheted through the pinball maze of traffic, I, nestled between my parents, my brain full of warring chemicals, felt, for the first time, safe.

BILL
WOODLANDS ASHA
NEUROPSYCHIATRY CENTRE

3-4-582, Opp. Provident Fund Building, Barkatpura,
Hyderabad - 500 027. Phone : 756 0918.

I.P. : No. ...*7.11.1*...............

No. : **2634**

Date : *20 - 10 - 2012*

Name of the Patient : ....*DAVID MACLEAN*..................

Address : .███████████████.....*au CAMPUS., TARANAKA, Hyd*

D.O.A. : ...*18 - 10 - 2012*....   D.O.D. : ...*20 - 10 - 2012*

PARTICULARS	AMOUNT Rs.	Ps.
REGISTRATION FEE	50	~
BED/ROOM CHARGES	500	~
MISCELLANEOUS CHARGES	50	~
ADMISSION CHARGES	50	~
MEDICINES	25	~
INJECTIONS	40	~
I.V.FLUIDS		
DISPOSABLES	120	~
INVESTIGATIONS	150	~
TREATMENT CHARGES *physician*	100	~
1. *counselling charges*	100	~
2. *nursing charges*	600	~
3. *consultation fees*	115	~
FOOD AND OTHER CHARGES		
(Rupees ...*one thousand nine*...	1900	~
hundred only............... only)		

For **WOODLANDS ASHA NEUROPSYCHIATRY CENTRE**

███████████

Authorised Signatory

We arrived at an apartment building. I had thought we were
going to their hotel, but my mom reminded me that we were
going to pack up my stuff. The building we pulled up to was
unfamiliar. Part of me still held on to the belief that my real
apartment was across from Mrs. Lee's guesthouse, that all of
this—the rickshaw ride, this apartment building—was part of
an elaborate ruse that had been prescribed by the doctors to
keep me from knowing the real truth of my life. Wasn't the fact
that nothing was familiar proof of this?

As we stepped out of the rickshaw, a scruffy man jumped up,
buttoned his blue shirt, and opened the gate for us. He called
me "Mr. Dah-wid." I shook his hand, and he gave me a key. We
got into the elevator, which was barely big enough for the three
of us, and ascended to the top floor. From there, I followed my
parents as they walked up a flight of stairs. This, then, was my
flat. The city rose up on all sides of us, giant rocky ridges to the
east of us and buildings clustered everywhere the eye roamed.
The city was enormous. My flat was at the center of the roof
behind a wrought-iron portico—sad and small, as if all it could
do was imagine how much bigger it could've been had someone given it the proper materials. My dad unlocked the portico
gate and then the flat's door, and . . . still nothing. I faked a little dumb show of recognition for my parents. I worried that if
I didn't start showing some progress they'd throw me back into
the asylum.

The place was painted pool-bottom blue. Oscillating desk
fans were bolted to the walls, only three fans for the entire
apartment, and the thick humid air barely stirred with all of
them on high. Dad went around opening windows and then
checked the small fridge for something to drink. There was a
tiny desk and a thin tiny bed. Everything was so small.

There was a small kitchen, a small bathroom, and a small

bedroom, all radiating off a normal-sized living room. There was a single chair, which my mom immediately sat down in. The chair was in front of a small desk, which held an open laptop. Bookshelves with sliding-glass fronts were inset into one wall and housed a row of books. Dostoyevsky, Thom Jones, Denis Johnson, a fat orange collected Cheever, Deborah Eisenberg, Alice Munro, Salman Rushdie, Barry Hannah, and William Dalrymple. All paperbacks, some with broken spines. Some creaked when I opened them. Some had writing in the margins, manic cramped scrawls with arrows and exclamation points that I recognized as my own, emphatically annotated appositives, ecstatic appreciations, careful observations. A copy of *The Great Gatsby* seemed especially attacked by the person who I had been. There didn't seem to be a line in the book that didn't have some kind of notation. The front pages, the ones that are normally left blank by the publisher, were crawling with quotes from the book pulled from their contexts and rewritten by me, or a me, maybe even *the* me. The one everyone kept expecting me to be again. Or was everyone trying to keep me from being that person again? I was getting everything mixed up. Were they upset that I wasn't who I had been? Or were they trying to prevent me from turning into that person again? The drug user. The one who ran around and did drugs with Christina.

The books sparked nothing, but I recognized the scrawling mess of my writing from my fever dreams with a ballpoint pen and the asylum's newspapers. It appeared that I was always trying to decipher something, even before I was insane. I slid the glass doors of the bookshelves closed.

A whining noise started, a metal-on-metal, crunching, high-vibrating E sharp. The room was full of it, like a mosquito with a megaphone. I checked my parents and was relieved to see that they were hearing it, too. Clasping my hands over my ears, I shouted, "What the hell is that?"

My mom tried to speak but held her mouth open until the noise stopped. "You wrote us about this. I think it's the elevator. The engine for it is mounted on top of the building." She rubbed the back of her neck as if she were trying to massage the sound out of her. "You found out about it after you signed the lease."

In the bedroom, I started loading my backpack, but I stopped almost immediately. I poked my head out into the living room.

"How long am I going to be with you all for? In Ohio."

My mom had been flipping through a magazine. "You're coming home, David."

I nodded. "I know I'm coming home. But how long did the doctors say I'd have to be there?"

"I think you should stay at home as long as you want," Mom said.

"But will I lose my funding? This grant that I have, will they yank my funding for this?"

"We're not having this conversation, are we?" My father leaned against the fridge. "Pack everything, David. You're not coming back."

My mom shot Dad a look, then turned to me. "We think you should stay at home and recuperate for as long as it takes to get this stuff out of your system."

"You're not coming back. What you just went through?" My dad was wearing a yellow polo shirt with a red gryphon on it and The MacLean Group written in script underneath. His eyes were granite obstructions. I took a picture.

"The cab's waiting. Pack everything. Let's go," he said.

"It's called a rickshaw," I corrected.

"Why is it waiting?" my mom asked.

"I didn't know if we could get another one, Sue." My dad's voice was becoming strained, yet it remained insistently calm.

"They're all over the place," Mom jabbed. "There were five just on this street."

"Our guy knows where we're going. And we have to get his prescriptions filled."

While my parents argued, I shoved a third of my things into

the green backpack, folded up my laptop, and tucked it into a padded slot in the messenger bag lying next to the desk. I opened the tiny drawer in the desk and found my passport and some rupees, along with a Leatherman and a picture of a woman who hadn't been in any of the pictures Betsy had sent me. This woman had curly hair, plump lips, and wide-set eyes that nearly vanished with her broad smile. I tucked her picture into the messenger bag.

I squatted between my parents and told them my plan. "If this grant thing is as competitive as you guys say it is, then I don't want to screw it up. I'm going to leave the bulk of my stuff here."

My father came toward where I was squatted. He looked like he was ready to take a swing at me.

"Dave, you've been through so damn much." It was at this moment that I noticed that his eyes were red, and only then did I start to realize just how much my parents had been through because of me. And here I was squatting like a camp counselor, explaining to them that I was going to do it all over again. My dad would rather punch me dead in the face than let me out of his sight again.

"It'll be okay, Dad. I'm on the proper medications now. I've gotten so much better in such a short time. I'll be back to normal very soon."

If this was all a ruse they were plotting to keep me from finding out my real life, I was going to play chicken with them. I'd tell them I was all right, that I was perfectly fine, and force them to tell me that I wasn't.

Mom sent Dad out with my prescriptions to find a pharmacy and to let him cool off for a bit. She and I went downstairs to see a neighbor who supposedly had visited me in the hospital, though I didn't remember him. Mom told me that he had been the one who had locked my apartment after the night watchman found my apartment wide open.

We took our shoes off and sat in a dark marble-tiled living room drinking sweet tea as the air was efficiently circulated by this man's giant ceiling fan. I could feel the cool air tickle the sweat on the underside of my arms. The man was a professor of history at the university down the block. His students called him Dr. ZoomZoom because he rode an old scooter to class.

My eyes wandered around his living room as Dr. ZoomZoom told my mother and me about the night I disappeared. I couldn't stop wondering if I'd been here before. Had he and I had tea in his cooled apartment before? Had I seen that picture, the one on the side table, of him with his arm around his teenaged daughter before? Did I know it was his daughter?

"So it was quite late, about ten only, when the watchman came and knocked on my door. I was in bed. I was afraid that there had been a break-in or some such trouble. The watchman said that Mr. David's door was wide open and there was music playing and the computer was on. He wanted to know if I knew where Mr. David was."

Dr. ZoomZoom was a thin gray-haired man with coarse black hair that grew out of his ears. He was clean-shaven except for a shadow of blue stubble on his neck.

"The watchman informed me that he had seen you leave the building at four p.m., and he mentioned that you had not been walking properly. He asked me if you had maybe some drinks now and then. I immediately told him that this was not the case and asked the watchman to take me up to your flat. I closed your laptop and unplugged it. The power surges quite frequently here. Turned off your lights and locked your door and placed one of my own padlocks on your gate. I gave the keys to the watchman and told him to alert me as soon as you came home. The next morning, when you had still not returned, I contacted Veda and Dr. Ramakrishnan, who I knew had been working as your advisor."

Dr. ZoomZoom took a sip of his tea before he continued. "Then the next I had heard of it was that USEFI had contacted Veda directly and let him know that you were in hospital, suffering from an allergic reaction. Veda then rounded up every American he knew to visit you. Even foreigners. He didn't want you to feel solely surrounded by Hyderabadis. This Richard chap nearly lived at your bedside. Veda found him in a library, believe it or not." He laughed and finished off his tea. "What was it, some malaria remedy or some such?"

"That's what the doctors think," my mom answered. "Lariam is what he was taking."

"What an awful thing. Does this kind of thing happen often?"

"It's rare," my mom said. "But the more we read about it, the worse it seems."

"They have these rings that you can burn at night, and they keep the mosquitoes away," Dr. ZoomZoom said. "Impregnated with some poison or some such, but better poison than what happened to you. All the thrashing you were doing. It was terrible."

"Thank you for visiting," I said, gritting my teeth.

"And with your memory loss. Think of it as a blessing. There are many things that I'd like to forget in my life."

I stood up. "My dad should be back any minute with the prescriptions. Thank you for the tea." Before Mom had even stood up, I was out the door.

I wasn't able to sit and hear the most nightmarish days of my life called a blessing. I lit up a cigarette, which unclenched the knot in my chest.

We left my flat and locked both the main door and the gate, and I put the keys in a special pocket in my messenger bag and zipped it closed. My mom walked to the edge of the roof and said, "Killer view up here. I love it."

I pulled out my camera.

Snap.

My dad and I sat at the bar of the Taj. Mom had fallen asleep
as soon as we got back to the hotel, so Dad had snapped off the
TV in the room and asked me if I'd go on a walk with him,
one that wouldn't involve leaving any air-conditioned areas. We
headed down to the lobby and then down some stairs, and there
was the Taj bar: a perfect replication of an English pub, with
gnarled wood on the walls, burgundy leather stools, and Bod-
dingtons and Guinness on tap. There was also a humidor built
into the wall with rows and rows of boxes of cigars.

The immaculate bartender slid two menus in front of us.
They were thick paper ones, the homemade kind of paper
where you can see the weave, and they didn't list any prices.
Dad ordered two scotches and two cigars. I found out later that
my parents had an agreement: one of them would be up to su-
pervise me at all times, and they would sleep in shifts.

The scotches came, his with two ice cubes, mine without.
Dad did the ordering, so I must have liked mine neat. The cigars came, too. The bartender clipped them for us, dressed the
edges, then patiently held matches for us as we puffed away.
Once both of our tips were red enough, the bartender slipped
out of sight.

I coughed immediately, having made the cigarette smoker's
mistake of taking the cigar smoke into my lungs.

I sniffed my glass. It smelled like Band-Aids. I dumped the
contents of it into my dad's glass.

"There's a new trick for you," my dad said, clapping me on
my back.

"It's all the medicines. I don't think I should drink on them."

He dropped his head. "You're probably right."

"The cigar's nice, though." The smoke was a different color
than my cigarettes, and it came out of us in clouds. The nicotine came on slower than my cigarettes, too. Smoking a cigar

was like having someone else wash your hair. You just gave in to the experience. There was a TV hanging from the ceiling in the corner of the bar. The volume was set too high, as if the machine anticipated a much bigger crowd than the two of us. We asked the bartender if he could turn it down, but he told us he wasn't allowed to. The TV pattered on in rapid-fire British English about a rugby match.

If my dad and I talked while he drank his double, I don't remember what we talked about. We sat in the loud and empty bar, smoking enormous cigars and feigning interest in foreign sports. It was the closest thing I'd had to normal in what felt like light-years. At some point I asked where I lived — was it in New Mexico or Ohio? — and we ended up drawing a map of my life on a bar napkin. My dad explained that I took a year off before college to work in the Honda assembly plant in Marysville, Ohio, and then I did college in Asheville, North Carolina, then a few months in Austin, Texas, and then four months backpacking in Sri Lanka and India, then briefly Raleigh, and then Chapel Hill, North Carolina, working low-wage jobs, and then Las Cruces, New Mexico, for graduate school and where I'd been for a year before I left for India again. The map we ended up with looked like one of the newspapers I'd marked up in the asylum, random arrows chasing each other in an attempt to assert some kind of order.

I was still catching things in my periphery, little cursive somersaults that disappeared as soon as I turned to face them. The world felt staged, some kind of kidproofed version of reality with all the sharp edges sanded off. After the neon explosions of my hallucinations, I felt like I was living in a consolation prize, a cheap imitation. I could almost hear the stagehands breathing behind the chintzy sets that I moved through. I had downed the Oleanz and Ativan pills as soon as we pulled away from the pharmacy, and they made the world feel shrink-wrapped. The Taj bar was a replica of an English pub, inside of a replica of a nizam's palace. The world felt facile and contrived, and I felt like I'd lost out. I'd seen a world made entirely of colors and connections, with tiny buzzing threads sewing everything together. I had been inside of a firework perpetually exploding, reaching out pyrotechnic tentacles into the outskirts of the universe, and now I was bolted into a world of satellite TV and drinks that smelled like bandages.

Dad finished his drink and signaled for the bill. It was 4,200 rupees, about $85. Two drinks and two cigars. My rent in Hyderabad was half that much. When my dad tells the story of traveling to India to pick up his amnesiac son from the mental institution, this bill is his punch line. It's one thing to have your son go crazy in a third-world country, but it's another thing entirely to spend $85 on two drinks and a couple of cigars.

We made the return trip up the stairs, the AC wrapping around us. Dad weaved a little. His jet lag combined with two scotches and a cigar was too much for him. We went and sat in two chairs outside by the pool. The air was hot and close, and I smoked cigarettes one after another. I relished the feel of the smoke in my lungs, the nicotine tickle that creeped up my skull. Dad dozed off in the chair, and I walked the lip of the pool, balancing on the edge, one foot in front of the other. The embers from my cigarette fell as I tapped it; they died with tiny sighs on the surface of the water.

The bed was strange. The comforter stiff and scratchy. The air unfamiliar. None of the shapes of the furniture in the dark matched up with anything I knew. Cold. I was melting in the coldness. My thoughts were pouring out of me and flooding the room. I was the end table. I was the cushioned chair. I was the telephone. I was the television. I had nerve endings sprayed throughout the room, and I wasn't sure how I could handle all of the data that was pouring in. My thoughts were puddling in the carpet near the doorway and sloshing down the hall. I was filling the entire building with myself and bursting the doors and coursing out into the street. I was the concentric circles of the gathering storm clouds. I was the pavement and the flower in the grass. I was the grass. I was the billboard and the black bird perched on it. I swallowed them both up before launching into the cosmos. The world in four dimensions had infected my synapses. I was drowning in the sensations of the entire universe.

I was one with everything, and I was terrified.

I screamed.

It wasn't much of a scream. It was enough, though, to wake up my mom and to get my dad sorting through the brown paper envelope with my pills. I was wet with sweat and breathing heavily, my asthmatic's rasp audible on the ends of each inhale. Mom held me tight and smoothed my hair as she shushed me into longer breaths. Dad slid the Oleanz and the Ativan into my mouth and coaxed a plastic bottle of water into my lips.

I was twenty-eight years old, a sweating man held intact in a hotel room by his mom and dad.

The next day we packed our things and left the hotel. Richard had come in the lobby just as we were leaving. He explained that he had been in the area while he shoved a packet of poems at me as my dad pushed past him. Richard really wanted to talk to me, to make sure I was coming back, and to give me some tips to guide my reading of his poems, but my dad held me close and dragged me past the man. It was like being famous.

Dr. Pat showed up as well. In the parking lot, she asked my parents to donate to her charity. She said that she'd been happy to help with my hospitalization and that she hoped my parents would be just as happy to help out those less fortunate. My dad gruffly said that we had a flight to catch and shoved our bags into the cab.

This was how I left India, with giant sunglasses covering half my face, huddled against my father, while a stranger who had appeared as God in my hallucinations shoved a ream of poetry under my arm. We were harried and hectic, Mom and Dad were arguing, and I was drugged to a calflike docility.

My father hated everything he saw in India. He was impatient and brusque with each person he met. It was hot. He hated the food. He hated Richard, who had visited me in the asylum and supplied me with cigarettes. To Dad, even Richard's poetry was guilty. He needed to reestablish a perimeter of safety to hold his son within. He would protect me this time. He was going to get me out of India. First he got me out of Woodlands, then he got me out of my flat, now he was getting me out of the hotel, and then he'd get me out of the cab, get me out of Hyderabad Begumpet Airport, get me out of the plane, get me out of the Mumbai domestic terminal, and get me out of the Mumbai airport. His trip was a list of places to unshackle me from. As we lifted off the tarmac in Mumbai and up the thirty thousand feet of airspace, heading north, he relaxed. He'd done his job. We were en route to Ohio. He'd saved his son.

On the plane, my parents kept at it: one watching over me while the other slept. I had the middle seat between them. One of their hands was always on me. Dad had brought a CD and a portable player for me. Apparently, I had hosted a classic country radio show on New Mexico State University's student station. It was called the *Baby Tyger Tri-Hour*, and I played a lot of Dolly Parton (when she was still with Porter Wagoner), Merle Haggard, George Jones, Hank Williams, Bob Wills, Buck Owens, Jimmie Rodgers, and Loretta Lynn. My on-air persona was intense. I opened with a fast-talking countrified accent and thanked people for tuning in so darned early in the morning, saying that I knew they had things to do, and I was there to put music on for them to do them things to. I told people I knew they were just cleaning the muck out of their eyes, buttering toast, walking the dog, getting out of the shower . . . and here I got really quiet and told people that if they were getting out of the shower, they should step up right next to their speakers. I told them that they should drop their towels and lean in close, and I would talk them dry, letting the vibrations of their speakers do the work.

Then I played a Conway Twitty song.

I was starting to learn who I was. There was a crazy person bellowing and talking nonsense between the songs. It was this person who was supposed to be me. I sat between my parents with my headphones on, listening. I knew that what I was listening to was at least a partial truth about myself. The person who was talking between the songs was me, but me while I was performing a part. The me who was on the plane listening to the me who was barking in the headphones decided that I needed to start acting like that DJ guy. I figured I'd try to understand the actor by performing one of his roles.

If I ever felt the jarring *thwock* of returning memories, it was sitting in that plane singing along to Loretta Lynn's "Rated X." I found out that I knew all the lyrics to every song that was played, but the lyrics didn't return in a flash; instead I knew them a half second before I was supposed to sing them.

The song lyrics I somehow knew, but I was still relying on what others told me about who I was. And the guy coming on in between songs, blathering nonsense about alien gods having scooped up the infant George Jones and touched his throat with their golden antennae—that guy wasn't helping at all.

On a stopover in Paris, I was presented to a Delta ticketing agent. There was a question about my return flight to India. Everything was booked from the week before Thanksgiving until mid-January. There were so many Indian students at universities in the United States, and they booked well in advance around the winter break, filling up every available flight.

Mom unzipped her leatherette datebook and flopped it open on the agent's desk. January stared me straight in the face. That meant the rest of October, then November, then December, then three weeks of January all in a town I had grown up in and didn't seem to have spent much time in since. Mom dragged her thumb across a row of boxes toward the end of the month and asked me to choose.

I reached up to the datebook and started flipping pages.

"When's the latest I can go back before the flights are all booked up?" I asked.

I was told November 18. My mom rubbed my neck and told me that I'd need more time than that.

"I'm just worried that they're going to pull my funding if I sit out too long," I told her. "It's competitive, right? There's probably a waiting list, and if I stay in the States until January, they'll give my money to someone else."

My mom told me that I should take this slowly, that I had been through a lot of trauma recently and that I'd need time to heal, that I should think in terms of injury, that I had been severely injured and needed time to let myself knit back together.

"But if it was the drugs that I was taking for malaria and those drugs are now out of my system, then I'm fine, right? It's not an injury, more like an allergy." Only hours before, I had been listening to the person purported to be me rave about how Wanda Jackson was a saint sent from Atlantis to salvage the aquariumed souls of the proletariat. Was it really the medica-

tion that sent me off the deep end, or had I always been standing a little too close to the edge?

What I did know for sure was that I was being handled gently. And I resented it. I fought against it. I was basically daring
my mom to tell me that the Lariam was just a theory.

My mom relented and let me book the flight, with the pro-
viso that I would need to be checked out by doctors and psychi-
atrists before I could return. Dad was out walking the terminal
and was thus prevented from voicing any objections.

On the flight over, listening to myself blather on the CD, I
had figured out that the quickest way back to mental health was
to appear sane. If I labored hard enough at the performance of
sanity, my insides would eventually come to reflect my exterior.
Faking sanity isn't as hard as it might seem.

You just have to shut up.

Our layover in Paris wasn't long enough for us to leave the airport, so with my dad trailing me, I wandered the duty-free shops. In a drawer in my apartment, I had found a wad of US dollars and stuffed them into my wallet before we left. Now I spent them freely, buying a carton of Gaulouises; a three-CD compendium of early US country music, its liner notes all written in French; and a duty-free hip-pocket-sized bottle of Johnnie Walker Black. I wandered into a café and bought a ham and cheese croissant and a cup of wine. I sat down at a tiny table and listened to the first CD. Milton Brown, Al Dexter, and the Carter Family. I knew none of these songs. I sniffed at the wine. It smelled like cigarettes and grease. I dumped it in the trash. The wood-glue feeling was sluicing back.

The airport waiting area had fluorescent light, which divided everyone into slices of gray flesh and fabric. The light was like a test tube over each individual. Even when people touched each other, I could tell they weren't really touching, that there were microns of distance between them. Each person was isolated, and they didn't even know it. The sadness of a life spent not knowing how alone you were broke open inside me. I picked at the seal of the carton of Gaulouises and searched for a smoking lounge.

Once we were airborne, my parents immediately went to sleep, despite their earlier efforts to sleep in shifts. They had endured the flight to India once already within the last four days, and before they could even deal with their jet lag, here they were again. The third leg of our trip was from Paris to New York. The inside of the plane was bleached with high-altitude sunshine, and the in-flight movie was too pretty and sad; I couldn't stomach gorgeous people crying. I unbuckled myself quietly, so as not to wake my parents, and walked to the back of the plane. The floor trembled with each step, and through it I could feel the tens of thousands of feet of pure nothingness between me and the middle of the Atlantic Ocean. I got to the back, interrupted the flight attendants, and peeled off a five-dollar bill for a tiny bottle of scotch and a cup of ice cubes. I poured the scotch over the cubes before I remembered that I didn't like it that way. I stood next to the plastic swoop of the wide rear door. Its half-yard metal bar was exposed, begging to be yanked open. Through its window, the size of a salad plate, I watched the clouds slide underneath us. The ocean looked like skin. I sniffed the scotch, and I smelled the plastic cup more than the alcohol. I took a careful sip. It had the cold metal taste of a watch battery. The movie must have ended because people were sliding the accordion doors of the bathrooms open and closed. I peeked down the aisle, and it was full of people stretching and shifting from foot to foot, all of them glassy-eyed and staring in my direction. I ducked back to my place next to the door and sipped at my drink, then crunched the ice, fixated by the clouds, the most ephemeral landscape, each moment an entirely new geography.

Ten minutes later, I chucked my plastic cup into the trash and made my way back to my seat. I was at the very back row when it happened again. The membrane between myself and the outside world dissolved, and I began to leak out. I grabbed an empty seat back to steady myself. My soul was quicksilver and sliding out through my feet. The trembling of the floor, the vibrations of the engine, the sunlight stabbing at me through the windows: I was being shaken loose of my body. Anyone who saw me probably thought I was drunk. And maybe I was. Scotch and Ativan is not a wise combination under any circumstances, and I was fresh from the asylum. I took some deep breaths, closed my eyes, and let the moment pass. The world reassembled itself, pieces falling into place inside me like a Tetris game. I opened my eyes and looked out on a world safe and separate from myself.

I made my way back to my seat and slid in between my sleeping parents, buckling my lap belt and placing my mother's hand back on my forearm.

Our next flight, from New York to Columbus, was a smear of turbulence. I held my dad's hand as the flight bucked up and down, trying hard not to imagine what was going on outside. The invisible wind currents that were buffeting us were dangerously close to my hallucinations, a world full of hidden forces and embedded codes that directed all of our activity. Somehow, everyone else was at peace with the notion of choppy air.

We landed and collected our bags from the conveyor belt, and outside we were met with the chilled night air of Columbus. A Cadillac, white and sleek as the belly of a shark, pulled up even with us, and its trunk popped open. A wide man lumbered out of the car and greeted us. His hair was as white as his car, and he wore a golf shirt tucked into his khakis. His enormous gut was bisected at the bottom—he actually had a cleft gut.

He shook my dad's hand and hugged my mom and called me Dave. We tossed our bags in his trunk, and Mom asked me if I was going to smoke in the car. I said yes, so I got the front seat. The man's name was Bud, and he had lived on our block almost as long as we had. Bud was the top management guy at the local hospital, and he chain-smoked Marlboro Lights. We pulled out of the airport, nearly the only car there at that time of night, and we sped down the deserted highway connecting Columbus with Delaware, the small town I grew up in. Bud and I both cracked our windows and lit cigarettes. He glanced at me every once in a while but didn't ask any questions beyond the status of our flights.

Dad conked out immediately, and Mom slid up between me and Bud, putting her elbows on the console as she talked a constant stream of plan-making to Bud. I was to visit the hospital and get a full checkup. Bud puffed away at his cigarette, laughably small in his enormous paw.

The streetlights pitched tents of orange glow up the length of I-71. The highway was a trembling marmalade wire stretched out through the undeveloped darkness of central Ohio. Bud and I smoked our way through it, his Marlboro Light cigarette butts the same white as his Cadillac, as his hair, as the edges of the ash we tapped into his overflowing ashtray. The cockpit of his luxury car glowed with lit-up information, the highway was lit up, and the spaces right in front of our faces were lit up intermittently, and we sucked that lit-upness into our lungs.

It took half an hour before we were deposited in front of our home. There was the off-ramp, the left turn, and the meandering stretch of blackness of State Route 36/37. I kept waiting for the moment where I'd recognize everything. Where the world would lock back into place and I'd become the person who belonged here. I waited for it like a dog waits for its walk. I wanted it to make sense. I wanted to feel the way I assumed everyone else felt: secure.

Bud stayed in the car while we unloaded our stuff and piled it onto the damp lawn. He powered down his window. I shook his paw, and he pulled me down close to him. He told me that I had no idea how much my dad loved me. Then he drove off.

Once inside, Mom asked if I would be okay by myself. I told her I was fine and asked her to show me how to turn on the TV. Dad punched the buttons and told me not to smoke in the house, and then they both went upstairs to bed.

I stood in the kitchen for a while. I was trying to assess if it was the kitchen from my hallucination, the one where all I had to do was go to the cupboard, pull down a pack of crackers, and say my trademarked slogan. The island in the center of the kitchen was there. But the kitchen was the wrong color, a burgundy rather than salmon. I opened up one of the cabinets and found bottles and bottles of liquor. I pulled one of the scotch bottles down and poured two fingers into a glass. Neat.

The house felt familiar, but I never experienced an avalanche of identity data. There was a dining room with wallpaper that had a fleur-de-lis pattern in red velvet. The extra-long living room where a previous owner had built an addition. My dad's office, painted a terrifying shade of red. It felt like a sham, more like a parade float than a house.

In the halls there were pictures of me. The glass on all of them was cold. The house was littered with versions of me. A giant one hung in the stairway (next to giant ones of my sisters): me, late teens, in a studio, dressed in khakis, a blue dress shirt, and an eyesore of a tie. I was standing there with the cuffs of my pants tucked and rolled at the ankle, bare feet, sleeves rolled up to my forearm, and my mouth wide open in a scream. There were the Sears portraits of my family in front of backdrops of bookshelves, my bad teeth and bowl-cut towhead front and center. On my mom's desk, right there next to her computer, there was a snapshot of me pretending to pee in a fountain and another one of me in a tight blouse and in a hoopskirt so tall, I must have been on stilts or on somebody's shoulders.

These were the images of me my mom kept close at hand? I couldn't tell who came out worse, me for having posed for the shots, or my mom for cherishing them. On her office wall was a blown-up photo in a heavy wooden frame: a very young, chubby-cheeked me in a white polyester suit sitting on a settee between my sisters, with a black-and-white dog stretched out on our laps. I didn't know the dog's name.

I could recognize these people in these photographs as me, but I felt a distance between us. They could all just as easily have been a chorus of doppelgangers. I felt myself slipping, worried that I'd never recover, that I'd be this wood-glue-filled piñata for the rest of my life. And then if I did recover, if I got everything back, who knew if it would happen again? How many times would I end up touring the exhibits of my curated self?

I sipped at my drink, which smelled like bandages, and turned on the TV. I sat on the couch and watched a documentary about an NBC weatherman's gastric bypass surgery. I watched the entire absurd show and drank the terrible drink, forcing myself to believe it was all totally normal.

PART THREE

He was many men and no man at all. He was a hysterical little bundle of possibilities that could never come true.

—Nelson Algren, The Man with the Golden Arm

I woke up on the couch with a dog's nose in my face. She was
large, and her tail was wagging so hard that her back legs frequently left the floor. Brown and black shaggy fur with a bear's face and what looked like black eyeliner surrounding each dark brown eye. She was easily ninety pounds, but she hopped up on top of me on the couch with the ecstatic disregard of a puppy. My mom popped her head into the living room. She was flushed.

"I tried to stop her, but when we were coming up the driveway, it was like she knew you were here." Mom petted the dog's massive head and shook her by the skin under her neck. "Sally loves her daddy, there's no doubt."

All ninety pounds of her squirmed onto my lap and began licking at my neck and face. A dog that looked like a bear. She recognized me and wanted to be petted. Life could be simple. I had spent the night watching my brain, waiting for it to spit sparks and malfunction again.

The wiry mess of fur and drool in my lap superseded those anxieties. *Pet the dog.* For brief moments, Sally the dog had the ability to force perspective and to make those spiraling thoughts the vestigial remains of a biochemical hiccup. *Pet the damn dog.*

"Marlee said Sally made a lot of friends at the kennel."

I pulled at her ears. "She's a good dog." I tried to remember adopting her, raising her, teaching her to poop outside and not to chew things, but the clearest images I got were like shadow puppets, shape and darkness, maybe a vet's office near some train tracks. She was big and wriggly, and I traded on her ecstatic recognition of me and shoved my face deep into her fur. She smelled like a cinnamon stick left all day at the beach: salt, spice, and sand. She writhed as I held her tightly, fighting me off in order to lick my face.

Bud got us an emergency appointment at the hospital. My first day back in the States, and we spent it in a series of waiting rooms, flipping through magazines with pictures of beautiful people whose names I couldn't place. At our first stop, we registered with the front desk. This first room had ceilings so high up that I got dizzy looking at them. A woman gave me a clipboard stacked with forms. Millions of little empty black lines. I puzzled over them and then pushed the stack of them into my mom's lap and went outside for a cigarette. On my way out I noticed that there were birds careening about the vaulted ceiling, tiny darting things that had nested near a skylight.

It was cold outside. I had worn only a hooded sweatshirt, and the wind made lighting my cigarette nearly impossible. I crouched down near the building to create a wind block. I felt the bricks humming, a deep vibration, against my shoulders. I sat down on a cold bench and tried to smoke as quickly as possible. The landscape was gray. A gray gas station, a gray strip mall with a gray Chinese fast food restaurant. The trees were gray, and the stoplight went from gray to gray to gray. The hospital was a shock of orange in the midst of this monochromatic batch of poor sketches. The wind blew hard and tried to shove the grayness into me. I relit my cigarette twice and tried not to think about the hospital vibrating behind me.

When I came back, there was a woman talking to my mom. The woman now had the clipboard, and as she talked to my mom she tapped a pen against her teeth. My mom stood up when she saw me, and I shook the woman's hand. We left the giant room and meandered through halls, a white on stainless steel maze, following the woman through doors that swooshed back and forth after we'd gone through them and into an elevator that opened with different doors than we had entered through. I wanted to ask if my mom had seen the birds but kept quiet because I didn't want to consider the implications if she hadn't.

I sat in a chair with one armrest and took off my sweatshirt, and a different woman asked me to make a fist, and then she stuck a needle in me, and I saw my blood bubbling out into a little glass tube. The woman removed that tube and put another one on the needle. My blood was sluggish inside those things. It barely trembled as she affixed labels on the tubes. There were tiny little bubbles at the top of each one. I wanted to ask her if that was normal. Maybe the fact that my blood had bubbles was the reason I couldn't remember high school.

After the blood was another length of twisting hallways to an X-ray, which involved a lead apron draped over my shoulders. They were seeing into my head. I wondered if they could see the birds flying around in there. Then there were more hallways, and I was injected with a contrast solution, and I emptied my pockets of anything metal. I was warned not to move before I was slid into an MRI tube. The machine was close to my nose, and it clanged and clanged as it turned me inside out. My body's secrets graphed onto a readout, stapled to a folder with my name on it.

Back in my street clothes, I sat with my sweatshirt in my lap as the doctor went over the results. I was exhausted. The doctor said that the MRI and the X-rays had both come up clean. There were no spots, clots, or lesions on my brain. He told me that it'd take a day or so for the blood work, but that the results indicated that the catalyst for the incident was nothing organic in my body, but rather a result of the Lariam. He then said what everyone said after they mentioned that drug's name: "The drug has a history of these kinds of side effects." We'd have to wait for the blood test results to make sure.

I didn't believe him. There was something definitely wrong. I wanted to tell him about the birds that I wasn't sure really existed. I wanted to tell him I could feel the building singing, that the world was monochromatic. I wanted to tell him that I was the liquid center of the universe and liable to dissipate into nothingness. I wanted to tell him to check again.

My mom asked him some questions. He said I should take a different malarial prophylaxis if I decided to return to India. My mom said I was acting strange before I had left for India, like "angry strange." The doctor nodded, but he didn't write it down. "Angry strange" was news to me. The first I'd heard of it. She said that I'd gotten very angry in my dad's office one night and had scared her. She pulled out a folder from her backpack. She had a whole spreadsheet typed out with every bit of information about the last couple of months. I found myself staring at the dates and numbers, and immediately looked away. She said that I'd arrived in India on September 22 and that I had complained of violent vomiting as of the twenty-fourth. I had written to her that the vomiting lasted about a week. Then she said that on October 12, I had walked into a Christian seminary in Hyderabad and told the priest that I had no idea who I was.

"This happened before?" I asked.

My mom told me that I'd gotten better after an hour, and then I'd called Veda, who came and picked me up.

"I remember being sick and this doctor coming to my apartment," I said. The scene of it blossomed in my mind: The doctor wore a stained blue shirt and hadn't shaved in a few days. He jabbed me in the upper arm with a syringe. "He gave me three injections, but he wouldn't tell me what was in them."

My mom told me that I'd been very nauseated and had been vomiting yellow stuff the night of the fourteenth. She said that Veda had told her that I'd been so delirious that when he came to visit I'd fallen out of bed and was sleeping on the floor.

The doctor said that all of this bolstered the theory that it was the Lariam. These events all seemed to be prodromal to the larger episode.

I had him define *prodromal* for me: ripples before the tsunami.

My mother's spreadsheet was littered with prodromata, little events building to larger and larger events. How could I be expected to understand what had happened to me when they kept using words I didn't know to describe it?

My mom explained the care I had gotten while I was in India. How I had started in one hospital and had been transferred to another. This was news to me. I thought I had been taken straight to the mental hospital.

"Which hospital did I punch the nurse in?" I asked, interrupting my mom's narrative.

"You punched a nurse?"

Out of all the things I couldn't remember, here was something no one else in the room had known. Mom and the doctor stared at me. Did it always feel like everyone was evaluating you? It was like their stares were another set of diagnostic machines. What were they seeing that I wasn't?

"I'm pretty sure I did," I said, trying to act casual. "Probably at the first hospital."

Mom resumed her narrative.

After the whole day spent in the hospital while jet-lagged and on Ativan and Oleanz, I felt like I'd been subjected to a pre-mortem taxidermy. I gathered up my sweatshirt, had the doctor sign a form from the State Department, shook his hand, and left.

It seems like I blinked my eyes and we were in Bud's office. He
was hunched behind his desk, and his office reeked of his chain-
smoking. While he and Mom talked about the tests and the antimalarial drug, I lit up a cigarette and added to the full ashtray.
Bud gave me a look like I was supposed to have asked permission.

Mom kept saying the drug's name, Lariam, how it had a history of doing things like this. Then she told Bud how weird I
was acting the week before I left for India, and how it all coin-
cided with my first dose of the drug. She now told Bud the story
of me sitting behind my dad's desk, so furious about a missed
fax that I threw a book across the room. She told Bud that she
knew something had been wrong but hadn't said anything.

Bud grimaced at me, and I felt like I was back at Mrs. Lee's
guesthouse, the kind of guy who terrorizes his mother, who
breaks his mother's heart. Just like a drug addict.

"What was the book?" I asked.

I sent out an e-mail to a list of friends and professors explaining what had happened. I told them that I was fine. I said that I had lost my memory and spent some time in a mental facility. I was fine now. I stressed that fact a few times. I had been lost but now was found. I was okay. I typed the sentence "I'm okay now" several times. I impressed on everyone that I was slowly recovering my memory and that I'd be back to normal very soon and that no one needed to worry. I needed everyone to know that I was fine. It was a funny story, really, when you thought about it, I wrote.

I was not okay, but it was easier to claim health than to explain why I felt like I was a step away from erasure at every moment or that I had begun to be suspicious of every mouthful of food. It was easier than admitting that I was suicidal.

I told them I was fine.

The next day the blood work came back. There was a high
level of allergen-specific immunoglobulin E. The doctor said
this was pretty good proof that what happened was the result of
the Lariam. He told me to continue taking the drugs I was prescribed by Dr. Chandra in India since they were what he'd have
prescribed me anyway. And that I'd need to find a replacement
malaria prophylactic for when I returned to India.

"There were no other drugs in my system?" I asked.

He explained that they didn't test for specific drugs so much
as try to gauge what my body's reaction had been, which to me
was like guessing what the stolen painting looked like by studying the spot where it had been hanging on the wall.

"More than likely, it was the Lariam," he said. "All we can
do is make deductions at this point. If there is a relapse of the
symptoms, we'll make a different diagnosis."

"It could happen again?" I asked.

"It's not going to happen again. I'm pretty sure it was the
Lariam."

"You said that those other times—with the vomiting and
not knowing who I was at the seminary—were all prodromal
events, right? Stuff leading up to the bigger event?"

He agreed.

"How do we know that all of this isn't prodromal to an even
bigger event? How do we know I'm done?"

"We don't," he said.

Malaria is one of the oldest, deadliest diseases on the planet. Its effects can be seen in the genetic record as far back as five hundred thousand years ago. In the history of infectious diseases, most of them burst on the scene and aggressively kill but then settle down and become less deadly over time. Killing your host turns out not to be the most viable strategy for a pathogen. Scientists call this process "diminished virulence."

Malaria is rare in that it has stayed as deadly as it has always been. Between 250 and 500 million people are infected with the disease each year, about 1 million fatally. According to Doctors Without Borders, reported malaria cases quadrupled between the years of 1982 and 1997 when compared to the data from 1962 to 1981. We've been battling malaria for our entire existence as Homo sapiens and even before. Our distant evolutionary relative the chimpanzee chews the leaves of the mululuza shrub, which has secondary properties that ease malarial symptoms.

For centuries, people in the Andes had known that the bark of the cinchona tree had properties that could combat malarial fevers. It was the Jesuit missionaries there who brought the powdered bark to Europe in the early part of the seventeenth century. The bark contained a molecule called quinine, and while it doesn't prevent people from being infected with the disease, it does soothe some of the disease's effects by poisoning the malarial parasite when it attacks hemoglobin in the blood. Quinine was humankind's best defense against malaria for centuries, although for a while it was a casualty of Catholic bashing, as people maligned it as "Jesuit's powder" and an agent of the Pope. Poor Oliver Cromwell died of malaria after publically dismissing quinine as Catholic hoodoo.

Malaria research and prevention were front-page news in the late nineteenth century as teams of scientists struggled to decipher the disease. In August 1897 in Bangalore, India, a town about four hundred miles south of Hyderabad, Dr. Ronald Ross and his Indian assistants discovered that mosquitoes transmitted malaria. Ross gets credit for the discovery even though a case can be made for an Italian team led by Giovanni Battista Grassi. Ross killed his mosquitoes for dissection by puffing some of his cigar smoke into their test tubes, which for them must have been like drowning in a bubble bath. This knowlege led to the second method of combat against malaria: fighting the mosquito. In the twentieth century, people did this as well as fight the parasite.

Malaria treatment factored heavily into World Wars I and II. Forests of cinchona trees around the world became important military assets. Prior to World War II, German scientists developed a synthetic version of quinine, but didn't use it because of its toxic side effects. Towards the end of the war, American scientists developed the same synthetic (corroborated by data found after the Allies captured Tunis). The Americans christened it chloroquine. This new drug was highly effective in preventing malaria, and people used it liberally, regardless of its toxic effects. Brazil even fortified table salt with it, which might sound extreme, but at the same time used similar logic as the United States' widespread use of antibiotics in livestock. The problem, though, was that malaria is tenacious. Soon chloroquine-resistant strains of malaria began to appear.

World War II also led to the development of a unique compound, dichlorodiphenyltrichloroethane, which when used as a pesticide had both remarkable staying and killing power. After a single application, this compound, referred to as DDT, could be present and effective as a pesticide in the soil for anywhere from days to decades. It was also easy to make and spread. Factories could churn this stuff out in mass quantities, and it didn't have the toxicity to humans that other contemporary pesticides had. In 1958, the United States (in a bill cointroduced by Senators John F. Kennedy and Hubert Humphrey) declared a global war on malaria, with an emphasis on killing the *Anopheles* mosquito, enlisting the new wonder pesticide known as DDT. The world's health organizations banded together and decided that with DDT they weren't just going to try to control malaria —they were going to wipe it from the face of the earth.

DDT's longevity (a half-life of eight years) and its widespread use for all sorts of insect control proved to be its downfall. Though malaria had been scrubbed out of the United States by 1951, American farmers dumped the pesticide on their fields. DDT breaks down into compounds that nestle into fat cells, taking years to metabolize. So when insects began developing immunity, those immune insects would then be eaten by predators, and DDT quickly crawled up the food chain, making its debut in the US milk supply in 1952. "All the things that we find sinister with DDT today—the fact that it killed everything it touched, and kept on killing everything it touched," Malcolm Gladwell writes in an essay in *The New Yorker*, "were precisely what made it so inspiring at the time." Public outcry about the pesticide ended its manufacture in all but two countries, China and India. Roughly a dozen countries still utilize the compound today, as well as people who still champion DDT and claim it should be mentioned in the same breath as life-saving, world-changing penicillin.

It's hard to overstate the effect malaria has had on human civilization. In her book *The Fever*, Sonia Shah points out that in the eighteenth century the British decided that due to the high rate of malaria outbreaks, sending their prisoners to Gambia was basically a death sentence, so instead they sent them to South Wales, Australia. She also notes that one out of fourteen human beings living today have genetic mutations that can be linked to the disease. Our history, genetically and culturally, is tied up with malaria.

There are two major kinds of malaria: *Plasmodium vivax* and *Plasmodium falciparum*, with the latter being more deadly.

The disease incubates in the guts of the *Anopheles* mosquito, and once it matures, it makes its way to the proboscis. The thing about this parasite is that through centuries of evolution it has developed systems that affect the behavior of its host in order to prolong its own life.

As malaria matures, the chemistry of the mosquito is altered, making it less aggressive, thereby protecting the host. When the parasite has matured and migrated to the proboscis, the mosquito's ability to create apyrase is decreased, which is the anticlotting agent that enables a mosquito to feed. Because of this reduction of apyrase, the host mosquito has to bite more often in order to be fully fed, providing the parasite with many more potential carriers. Over millions of years of evolution, a series of complex chemical occurrences has developed, facilitating this perfect relationship between host and parasite. The host becomes the servant to the thing living inside of it. Its behavior is changed to make it a more effective vehicle for the parasite.

When an infected mosquito bites you, thirty or forty parasites slide into your bloodstream, set up shop in your liver, and get busy multiplying. After a week to fourteen days (although there are reported cases where the infected person can go months before showing any symptoms), the disease hits a critical mass and releases into the bloodstream again, going after the hemoglobin, attaching to their studded proteins, and eating them from the inside. The fever occurs when the parasites excrete their waste after feasting on your hemoglobin.

The first symptom you'll generally feel is a full-body chill. Your body will begin to shiver violently as the autonomic system tries to warm itself. Then the fever begins. This transition from chills to fever is the malarial trademark. The fever can rise to 106 degrees, boiling you from inside. Untreated, you will likely become anemic, slip into a coma, or die. The disease exhausts you, forcing you to remain prone and therefore easier bait for other mosquitoes. It's a downward spiral from here, as the disease has your body release a pheromone that makes you a more appealing target for mosquitoes. During this time mosquitoes can attack you and spread the disease to as many as one hundred other people.

Malaria separates the native from the visitor. The native tends to harbor an immunity to the disease as a result of hundreds of thousands of years of evolution. (My favorite evolutionary adaptation is the production of blood cells that don't have studded proteins. The smooth blood cells don't give the parasite anyplace to hold onto, and malaria slides off these cells like they were made of glass to then be dealt with by the immune system. When this genetic trait is doubled in a person, it produces sickle-cell disease.) When we travel to places we don't belong, even our blood is conspicuous.

My sister Betsy was at a Jungle Brothers show at a club in Maine in the spring of 1993. She was on a date with this guy named Eric, whom she had been seeing for almost the entire spring. She fainted in the middle of the show. Though my sister is six foot one, Eric picked her up and rushed her home. The next day Betsy went to a health clinic, where she told them she thought she had malaria. She had been in Kenya, but that was eighteen months earlier. Travelers are tentatively cleared for malaria-related illness three months after their return and by twelve months are fully cleared from the disease. The clinic disregarded her hypothesis, sent her home, and told her to return if she had a fever. The next day, Eric brought her back with a raging fever. A blood smear discovered the scribbles of malaria in her cells. The disease had incubated in my sister for a year and a half before emerging to ruin her date. For eighteen months she had walked around with one of the most deadly diseases lodged in her liver. Her heart bent toward Eric while her liver was full of parasites. Betsy got better, and Eric was at her bedside the entire time.

Betsy has a hard time showing weakness. She had played basketball in high school and then in college was a walk-on center. She ended up in the starting lineup. She believes that strength is a virtue, and because of this trait she can be a person easy to admire but difficult to feel close to. But when this disease came, she was gathered in Eric's arms and nursed back to health with him nearby.

When malaria is inside the mosquito, it manipulates its behavior. I wonder what effects the parasite had on Betsy's behavior. She played half a season of Division III ball while infected, and after a lifetime of indifferent dating, she fell in love while infected.

She dated Eric for ten years and then married him a month before I left for India. I can't remember the ceremony because I was taking a drug to prevent malaria.

After being home for three days, I talked to Anne on the phone. According to my parents and e-mails in my in-box, she was my girlfriend. Her voice was faintly familiar, like the smell of the car heater the first time you turn it on in the fall. It was a sweet voice, with the vowels flattened by an upper Wisconsin childhood. I could not remember anything about our relationship, but the picture I pulled out of my desk back in my apartment in Hyderabad, the one with the girl with the wide-set eyes and sweet smile, turned out to be of her.

I had already learned how to fake recognition. It was a survival tactic in my tiny hometown, when walking across the street to the grocery store would involve running into eight people who knew me. The skills I used to get by on a daily basis are the ones used by any con man. I let the other person lead the conversation, and I agreed with whatever was said, adding bland rejoinders, if necessary. It was like having a conversation when your face is full of new stitches and you have to be careful not to split any of them with your emotions. I was a newly stitched-together doll of myself, and thanks to the Oleanz and Ativan, full of cotton batting.

The conversation with Anne felt like a spectator sport. I couldn't believe she wasn't aware of how far away I was. She was having a conversation with this effigy of me. She oozed care and goodwill, and I was suspicious of it. Was this all some prank that she was pulling? She said that she wanted to come visit. I listened as she said that Halloween weekend would work the best and that she'd found a good deal from El Paso—and she was supposed to fly into Columbus, right?

It was like watching a filmstrip in science class. What was on the screen was important and I'd be responsible for the information later, but at that moment it was all blurry, distant, and two-dimensional to me. I had a day left on the Oleanz, so I chalked up my grogginess and distance to the drug. I watched myself speak into the phone. I heard myself tell Anne that I'd be better soon, that I was looking forward to her visit, that if she sent me her details I'd pick her up at the airport. She told me that she'd been so worried about me. She told me that she loved me. I had a photograph of her, some e-mails between us, and this phone call. Whatever she had been to the old me, I couldn't really make out. In all of the pictures from my sister's wedding that August, Anne was in none of them. If I was in love with this girl, why hadn't she been there? But here I was now, in trouble, and she sounded so worried about me, genuinely worried about my well-being.

I told her that I loved her, too.

Inside of me something felt like it was shifting, the way the wineglass must feel the moment right before the magician yanks the tablecloth from underneath it.

The first response I received from my mass e-mail was from a former professor. He told me he was sorry to hear about my incredible predicament and that, in unrelated news, he had been going downstairs to do laundry and tripped on the stairs, hit his head, and now had superpowers. He said that he'd been flying around his living room all afternoon and that he wished me the best. It was clear he had thought I was kidding.

Whoever I was before the incident, I had been a smart-ass. It looked like I was exactly the kind of guy who'd make up something like this as a prank. Just getting people to believe that this had actually had happened was going to be difficult. Whoever I had been was now preventing me from explaining to my friends what had happened to me. He was a guy who prided himself on never giving a straight answer.

It was like I was running down a long hallway, at the end of which was a figure, this person who I had been, who I was supposed to become again, but the hallway was dark, and I kept stumbling over things that the old me had placed in my way. I could hear him laughing as I tripped and fell down in the darkness.

Jon and Melissa, my best friends according to my parents, called me. There were loads of pictures of the three of us in a photo album I'd found in the attic. I had been the best man at their wedding. Jon and I had been roommates in college, and later I had lived with the two of them in a couple of places throughout North Carolina, one of which was condemned as soon as we moved out. I was remembering blips of things. Jon was tall, whip thin with a concave chest, while Melissa had wavy light brown hair and a nose ring. I remembered dancing with her at a wedding. She'd taken off her heels so she wouldn't be taller than me, and she held them in her hands, which were around my neck. I couldn't remember addresses we'd lived at or what either of them did for a living, but I remembered the way her heels bounced and clicked against my shoulders as we danced.

They were calling together, each on a separate extension in their condo in Raleigh. I was sitting in my dad's bloodred office with the giant round red rug, the same room where I was said to have thrown a book and scared Mom. Dad's desk took up most of the room. He had a great cushy leather chair, and I leaned way way back and propped my feet up on his desk.

"Just the sort of thing to happen to me, right?"

They laughed. "As soon as we heard, the first thing we thought of was Bob Trace," Melissa said. "I told Jon that one of us is next."

"What happened to Bob?" I asked.

"You don't remember that? Bob Trace was in Berlin and took a bunch of acid and ended up running around the city naked until the police caught him and put him in a German mental hospital." Jon cleared his throat loudly as he spoke.

"*Der Krankenhaus,*" Melissa said in a German accent.

"But I didn't take any acid in India." I wanted to word it like a statement, but it came out like a question.

"No. But we all did some crazy crazy stuff back in college. I wonder if this is all like a long-term flashback or something."

"I wouldn't have taken acid in India, right? I don't do that stuff anymore."

"No. No. You haven't done anything like that for years," Melissa quickly said. "You should call Bob Trace, though. It's crazy that this happened to both of you."

She gave me Bob's number, which I penciled on the back of an envelope.

The phone call ended with Melissa telling me that I sounded okay, that there was probably no need for them to come and check up on me. I desperately wanted them to visit, but to say as much would have been to refute her claim that I was okay, and I needed to be okay. I had to be okay.

I agreed with her. Yes, I was fine. They wouldn't have any fun on their visit anyway. All I did all day was sleep.

I hung up the phone and tore the envelope to shreds.

Combining Oleanz, Ativan, and jet lag is a good way to average roughly seven conscious hours a day. I slept like a teenager. I woke up to a stack of phone messages from concerned friends. The hours I was awake I spent rereading all of my e-mails up to October 16, the day I walked out of my apartment with the door wide open. I also scoured my computer for pictures of the last couple of years. I was chasing myself, tracking my whereabouts, hoping that I could reconstruct enough of a working resemblance to that old self to slip back into. It was like building a plane while flying it.

There were more pictures of Anne, as well as more pictures of the woman with the amazing nose. She was more exotic than Anne. There was nothing of the Midwest in her features. My mom said I had dated her for two years and that things had ended very badly. I vaguely remembered arguing with her in a restaurant. Memories were coming back, but they were still shadowy, and I had to work incredibly hard to recall what should have been the simplest thing: her name was Ariel.

I got an e-mail from a cousin. She said she was so sorry about what had happened to me and that she'd had such a great time talking with me at Betsy's wedding; she was happy that we'd reconnected and really wanted to read some of my work.

I wrote her back. I punched letters on the keyboard and told this woman how lost I felt, how messed up I was, how I was wiped clean and struggling to remember anything, much less the people I talked to at weddings. I told her about *prodromal*, Lariam, and *pleomorphic*, this stupid vocabulary that I was thrust into. I told her that the doctor couldn't tell me for sure if it all was going to happen again at any moment. This was rat-crap bottom, and I was grasping at anything to keep from slipping into total paranoia. I was still kind of convinced that God hated me and was disappointed in how stupid I was. The e-mail

kept scrolling farther and farther down the page in a single un-
broken paragraph of rage and fear and loneliness addressed to a
woman I couldn't have picked out of a police lineup of three.

I erased it and wrote that of course I remembered talking
with her at the wedding and thanked her for her concern. *It was
wild there for a while,* I typed, but I was fine now. *Perfectly fine.*

I sent it off, got my cigarettes and a scotch, and sat outside in
the cold, watching the stars, begging them not to move.

A week after I'd come back to the States, ten days from my release from Woodlands, I slid into a padded chair across from a local therapist. It was the day before Halloween, and his office felt like a paneled womb. The man across from me had a beard and a clipboard and a sweater. I handed him the form the State Department had faxed me. Dad had given up the fax machine in his office after I left for India, claiming that he was printing out more junk mail than anything else, so all of our faxes came and went through the grocery store across the street from my parents' house. All of the documents of the current and future prospects of my sanity were slid across a counter covered in merchandise branded with my high school's mascot and faxed from a machine wedged between the cigarettes and lottery tickets. The therapist I was seeing was a friend of a friend of my mom's. He'd come recommended. He came out from behind his desk, shook my hand, and sat in the other padded wingback chair.

Before we began, he said, "I am going to say three things to you: White horse. Purple rose. Blue fountain."

For the next forty-five minutes, I told him my story. He was the first person with whom I shared the uncut version. Within five minutes he nudged a box of Kleenex toward me. I told him about the train station and about being a drug addict and not knowing who my girlfriend was and how everything felt queer and distant and lonely and infused with failure, because part of me still believed if I'd answered God/Jim Henson's question correctly in my hallucination, I wouldn't be stuck in this stupid world.

Forty minutes later, when I had a fistful of damp shredded tissue in my hands and my legs pulled up onto the chair with me, as I sobbed and told him how I felt scooped out like a jacko'-lantern, he interrupted and asked me to repeat the three things he had said at the beginning of the session.

I said, "White horse, purple rose, blue fountain."

He checked a box on the form and scribbled his signature on it. He handed it to me and said that there was nothing wrong with my memory and that I'd be fine to return to India.

Anne left a message for me with my parents and then sent me an e-mail to make sure I had gotten the message and to call her immediately. I wrote her that I wasn't very good at the phone right now, and my sleep schedule was way off, and that we could do what we needed to do on e-mail, right? I ended the e-mail by telling her how much I loved her. These words felt required. I also needed her. She was a key to this other life of mine.

This was at four a.m. Wide awake, I alternated time on the computer with time sitting in the backyard, where I drank scotch and smoked while Sally sniffed the frost-brittle grass. I also exchanged e-mails with Geeta. I had a picture of her on my laptop from the Fulbright orientation. She was gorgeous, Indian American, another woman who didn't have an ounce of the Midwest in her. From our previous e-mails I had deduced that I'd been meaning to visit her in Goa. These plans might have been the reason I ended up at the train station on October 17. I'm not sure what I had meant to do with her in Goa, but from looking at the pictures of Geeta and from the flirty tone of our e-mails, it was pretty clear I had never told her about Anne.

There were still many hours of my life that were completely wiped from my record. My friends and family could help fill in only so much. Those hours from four p.m. on the sixteenth till I woke up in the train station were a mystery. Nobody I knew had witnessed them. Why had I left my apartment? Where the hell had I slept? What was it about Geeta that had drawn me to the train station and not anywhere else in Hyderabad?

Geeta wrote, making sure I knew that the invitation was still open whenever I needed it. That I should take all the time recovering that I needed, but she really missed me, as well. She kept saying that she was so sorry for what had happened to me and that she'd gone crazy with worry. Then she'd tell me about her bikini and how she needed me to be her "white husband" in order for her to wear it at the beach. I e-mailed back that I'd be happy to be her white husband, and if it would make her more comfortable, I'd wear a bikini as well. Our e-mails ended with x's and o's, which I fretted over till dawn, not knowing if either of us meant them in earnest.

I took Sally outside so I could smoke and drink while she sniffed and peed. I'd just made promises of a sort to two different women, one who was attached to the old me and the other whom this new me was attracted to. It would be best to be honest, but I had no idea what the truth was other than the fact that I could walk by both of these women in the grocery store and not have a clue who they were.

Another response from my mass e-mail, this one from Lizzy,
a fellow Fulbrighter, began with the line "I'm going to be very
genuine and act on the assumption that you are not making this
up" and ended with the line "Good luck to you. And if you're
making this up, you are a big jerk."

I could only have known this woman for a maximum of two
months, yet she knew that I was capable of creating fabrications
and passing them off as truth.

I asked my dad about these e-mails — the one from my professor and now this one from a woman I had just met. Why
were people reluctant to believe me?

He told me a story about when I let the front yard's grass
grow really long, played hooky from high school, and then
mowed circles in it. I called the local paper and claimed that
aliens had come in the middle of the night and made them. I
begged them to send a photographer over right away.

Then he told me a story about how I had written a review
in my college's newspaper of a new album by this band from
Ohio. The only thing was that the band never existed. The college activities director wrote to the fake address I had supplied
at the bottom of the article to inquire about booking the band
for the school's spring festival. The address was a friend's, and
I had kept up a correspondence with the activities director for
months, apologizing for delays, asking if the demo I had sent
had gotten through, cursing the mail system when she wrote
back to say she had received nothing.

"You do have a history of this kind of thing," my dad said.

"I'm not doing it now, am I?" I asked. "This really happened,
right?"

With the Oleanz done, I tapered myself off the Ativan, keeping a couple in reserve. My mom had pulled out every single photo album that she had, and in the afternoons, while my parents were at work, I'd pore over them. There were almost no pictures that didn't have me mugging for the camera: bugging my eyes out, twisting my lips; I was predictably off-kilter in nearly every picture.

My memory was coming back, but at a glacial pace. I'd remember parts of my experience at tiny little Warren Wilson College, but not all of it, which drove me frantic. I'd remember a walk I'd take every morning from the cafeteria to the organic garden where I worked, but I couldn't remember what I did there. I'd remember that month I had a mania for ice cream cakes that Dairy Queen sold two for one on Tuesdays, and how I'd have to cut the cakes into thirds and stack them so that they'd fit in the dorm fridge, but I couldn't remember who I ate them with.

My years living in Chapel Hill were hazier. During these years, I was bumming around, working a dozen different low-paying jobs, guiding outdoor trips in western North Carolina during the fall and spring. I had a red Honda motorcycle. I had a belief that when people's memories were working properly, they remembered everything. And these shards of memory I felt were evidence of how damaged I was, instead of realizing that that's the way most everyone remembers.

I could remember Ariel a little better. I remembered that we lived together for a time, but not at the end of our relationship, which was "pretty dramatic" according to what Jon and Mel had said.

I found a book that Ariel had made for me. It had a copper cover with SO FAR stamped on it. Inside were collages of pictures, diary entries, drawings, and ticket stubs. On the second to last page was a list surrounded by pictures of the two of us. The list was titled "Clearly . . ." and took up the whole page:

Shower curtain hugs, tagged pumpkin seeds, banana pudding skin, white heat dancing, best western inn, NYC rooftop glimpses, campers quarters, pushed together beds, lost keys, tekken, air hockey, banana protein shake car spills, the big sleep, big man singing on the metro while we doze, rose garden what are we doing conv., dirty pullout couch, late night bed talks, walking into a bare room w/ Joni Mitchell playing looking down at a beautiful boy, cheesy yes, a sweet call from mountain man boyfriend at chip's party, jon's b-day party, all around stimulating interaction, yes, I'd like to think.

It was code, some sort of relationship code, embedded throughout with intimate detail but impenetrable to me. It was a fossil, and I stared at it, trying to imagine what it looked like when it was alive and roaming the earth. Hallucinating in the asylum, I thrashed against the straps, trying to figure out the quatrain that Jim Henson asked for. I had gotten half of it — "If you can't understand that the world was made by Jim Henson in a studio in Burbank, then fuck masala dosa . . ." — but the other half had eluded me, so I never got to pass into the next level of experience.

This list in front of me was another opportunity for fresh failure. Why couldn't my brain work? Why was everything a jumble inside me? I was slowly getting memories back, but I had no idea what they meant. They were just scraps of evidence with no sense of the criminal they pertained to, or the crime.

I devised a system. I'd study a picture, catalog who I was with, where it was taken, and what I was wearing, and then I'd imagine a scene where all this had taken place. I'd imagine the jokes that were told; I'd imagine the moments of fond exchange; I'd imagine the slight enmities that were running below the surface. I'd filter it all through the character of me that I'd inferred from reading my e-mails. A goofy, self-deprecating loudmouth who preferred to say outlandish things rather than attempt real conversation, but who, in rare moments, could provide intimate and earnest observations about another person's well-being. I dubbed this voice onto the action captured in the photographs.

Using this system, I placed myself at my sister Betsy's wedding the previous August. I was dressed in a kilt and emptying garbage cans into a Dumpster. I placed myself in a parking lot in Asheville, wrestling with Ariel on the asphalt. I had located a stack of photographs in a box that I'd put in my parents' attic, more recent photographs, and through them I placed myself in the house I had been renting in New Mexico, playing darts in the living room with Anne.

Anne. It seemed I was fond of her, but distant. In the photographs, you could sense the distance. The way I always appeared to be pulling away from her, looking somewhere else. She had a tangle of brown curls that fell to her shoulders. Her eyes were wide set, and her nose was cute and a little upturned. There were pictures of us at a wedding. She in a purple sheath dress and a straw bucket hat, I in the suit that had been hanging in my closet in the flat in India. She stared straight at the camera and smiled, her arm stretched out and resting on my shoulders. I was talking to someone off camera. I didn't like that when I looked at the picture I immediately thought about how much more attractive Geeta was.

It seemed like I was always pulling away from women who

liked me. Even at that moment, I was doing it. Anne was coming to see me, the only one of all my friends to do so, and all I
could do was write to another woman. I wanted to not be the
guy in the pictures. I wanted to be happy with someone who
wanted to make me happy. I made a vow to stop writing Geeta.
I was going to love Anne.

The pictures weren't always helpful. The ones from college were odd. I was in green-feathered pants and a sleeveless sailor blouse with two baby-doll heads taped over my chest. There was a series of black-and-white art shots of me wrapped in a satin blanket standing by a lake; the other shots revealed that underneath the blanket I was wearing tights and an aluminum-foil codpiece. I asked my dad about these pictures. He wondered if there were any of me in pink saran wrap.

"It was your Halloween costume one year," he said.

"What else was I wearing?"

"Just the saran wrap." He sighed and said, "You've always been unique."

My brother-in-law's parents worked at the National Institutes of Health. When I turned up in India with no memory, they did a literature review of Lariam, whose generic name is mefloquine. As I recovered in Ohio, my inbox filled up with the studies they forwarded me. I didn't read any of them. I didn't want to know the possible scope of my condition. To read them would have been to know, and to know would make me acknowledge that I was still a mess and not making the best choices for myself recovery-wise. Even a glance at the studies about the neurotoxicity of Lariam revealed that alcohol was the worst thing (other than more Lariam) I could have been putting into my body. And I wasn't about to stop drinking.

Malaria evolves quickly, sliding past our miracle cures in a generation or two. During the Vietnam War, when there were outbreaks of chloroquine-resistant malaria among the US troops, the military scientists at Walter Reed Army Institute of Research (WRAIR) set about testing 250,000 new compounds to fight against malaria. The WRAIR is named after Walter Reed, the military scientist who demonstrated that mosquitoes were the vector for transmitting yellow fever. His work was used to control mosquitoes in Panama, allowing the construction of the Panama Canal. In their quest to test their hypothesis that mosquitoes transmitted yellow fever, Walter Reed's colleagues, Dr. Jesse Lazear and Dr. James Carroll, infected themselves with the disease and charted their symptoms. The vector was proved, but at a cost. Lazear succumbed to the disease on September 25, 1900.

In 1971, the WRAIR found that compound number 142,490 (which had been first synthesized in 1969) was able to prevent the malarial parasite from developing in the lab, but they didn't know exactly how it worked. The understanding of why and how a drug works isn't necessary for its commercial availability. If it works, it gets a price tag.

Conceived initially as a treatment for people already infected with malaria, the drug was tested on inmates at the Joliet Correctional Center in Illinois. Of course, these prisoners first needed to be inoculated with malaria to see if the drug would work. After this testing showed the efficacy of the drug as a cure for malaria, it was then developed in conjunction with Hoffmann–La Roche, a Swiss pharmaceutical company. This was the first time a public-private venture was undertaken in regard to the development of a drug.

The drug began to be marketed and used as a prophylaxis against malaria, something a patient could take to prevent infection, but never received the randomized double-blind study normally required of a drug for FDA approval. When the drug became available in 1989, Lariam was widely adopted by Peace Corps volunteers in malarial regions. An observational study was set up to record the PCV's experience with the drug. These volunteers adjusted their doses as they saw fit. Some experienced troubling side effects, like insomnia and incredibly vivid dreams, and stopped taking the drug entirely. This information about these varied doses didn't flow upward to those doing the observational study. The Peace Corps volunteers had received the drugs; the study assumed that the drugs were taken as directed. They were not.

The volunteers were prescribed 250 milligrams every two weeks. When some still came down with malaria (most likely those who had altered their doses or had ceased taking the drug), the recommended dosage was bumped up to 250 milligrams every week.

It's important to note here that Lariam seemed to fit the bill as a "miracle drug." It was effective and convenient. It only needed to be taken once a week. It seemed like it was the perfect thing with which to combat malaria. And it was, if you ignored all the reports of extreme reactions to the drug people were having. These reactions were easily dismissed as the reactions people have when they travel to stressful places.

In 2001, a randomized double-blind study was finally conducted on Lariam by a group of scientists based in the Netherlands. Sixty-seven percent of people who took Lariam experienced one or more adverse effects, and six percent had reactions so severe they required medical attention.

And even that study doesn't get to the bottom of how bad Lariam can be.

It hadn't taken as much effort as I thought it should have to convince my parents that I was okay to drive to the airport. They'd drawn a map and told me I should allow about an hour for the drive. The steering wheel felt enormous, and I questioned their decision the entire way there.

Inside the terminal, a crush of people. Some sort of tournament was going on in Columbus, and the airport was filled with lanky, muscular young men and women traveling in groups, each defined by the colors of his or her warm-ups. They all looked slightly sick in the airport's fluorescent light. They were on the outer edges of puberty and had grown-up muscular bodies stretched tight within teenagers' skins, as though their bodies were elaborate machines they wound up in and hadn't gotten the hang of yet. They bustled with an energy that disturbed me. If they channeled it all at once, they could tear the place apart and no one could do anything about it. I was Red Riding Hood in a forest of wolves.

I was never going to find Anne in this mob. In this age of cell phones, we forget how we used to pick people up at the airport. It was always a small miracle when a pickup happened. I made my way down to the baggage claim and found more troops of warm-up-wearing giants. Men and women carrying clipboards shouted things at their groups. The normal travelers were dumpy squat things in comparison, wrapped in blacks and grays and browns, wedging themselves between the groups to haul their black, gray, and brown luggage off the conveyor belts. Having resigned myself to finding the intercom person, I bumped into a woman who was bent over, adjusting the straps on her backpack. In the middle of my apology, she stood up and hugged me. Then she kissed me in the midst of all of those strangers.

This is how I found Anne.

Anne and I walked, holding hands, up to the roof of the parking garage. She was talking about a presentation she had had to give on Foucault and how she was never sure how to pronounce his name, so she had attempted to give the presentation without ever once saying it. I was amazed at how easily she let herself be driven by me, as if my competency as a driver was never in question by anyone. I could feel the thrum of the tires on the highway through my palms. She listed people who wished me well and who were eager to hear exactly how I was doing.

Anne started talking about the magazine she edited and told little anecdotes about the crazy things crazy people write, and how she was always at her most diplomatic with the crazies.

I watched her out of the corner of my eye as I drove. I tried not to think of Geeta and felt guilty that I had to make that effort.

At home, Sally immediately recognized Anne and squirmed in ecstasy at her arrival. My parents hugged Anne and welcomed her, my dad taking her backpack and putting it in my room while my mom made her a cup of tea. We ordered pizza that night from a place my mom called my favorite and watched a basketball game. Anne told stories about growing up watching Packers games in her tiny town in Door County, Wisconsin. It all was so easy, to sit and talk and watch basketball. Why had I left all this to go to the other side of the planet? The ease of the night was strange: everyone leaning forward to hear each other precisely, everyone eager and pleasant. I had three scotches. Anne talked about the magazine some more, and then we went to bed.

Anne and I undressed on opposite sides of the bed and climbed in with the lights off, red electric light coming in the window from the grocery store sign across the street and spilling onto the carpet. I put my arm around Anne. She said we didn't have to do anything if I didn't feel up to it, that she was happy just to be there and see that I was okay.

"I'm fine," I said. "Scared and disoriented, I guess. But okay."

She kissed my chest and said that she loved me. I said that I loved her, and we fell asleep, her warmth next to me strange and wonderful.

I woke up in the middle of the night, and the world wasn't behaving. The dark corners of the room were full of bats, and if I moved, they would all wake up and attack me. I stood paralyzed in the middle of the room. I moaned. The room felt disconnected from the house, and it was floating in space. I could hardly breathe. My chest was collapsing and crushing my lungs. The room changed again and again. The black places in the room were liquid antimatter and would freeze me if I touched them. The black places in the room were sentient malignant masses of teeth. The black places in the room changed over and over into different things, but they were always malicious and angry. The only safe place was the one strip of carpet lit up from the grocery store's sign. I stood locked on that strip, whining and trembling. Anne jumped up, put her hands on my shoulders, and guided me back to bed, pushing my hair back and telling me that it was all going to be okay. Scared and ashamed, I blubbered and told Anne she was the only thing holding me back from all the nothingness.

The next morning, my mom took us across the street to have breakfast with Dad, who'd been sipping coffee and holding court at the grocery store's café for at least an hour. Back in Ohio, he was bombastic and charismatic and seemed to know everyone. When he talked to someone, it seemed like he was talking to everyone in earshot at the same time. Not wasting that charm talking to individuals, he broadcast it to the entire room. We sat at his booth, and when the waitress came over, he chatted with her about her softball team and her recent knee injury. She was a bubbly woman in her late twenties, and she took orders from Mom and from Anne, and then snatched my menu away from me and told me that she knew what I wanted, reciting exactly what I was going to order.

"She's good," my dad crowed.

After the waitress left, Mom laid out a plan for us for that day. We'd take Sally and go out hiking at the dam, and then maybe go see a movie and go out for dinner. Still reeling from the night before, I clutched Anne throughout breakfast. Mom asked if it was okay that we spend the day with them. Anne told my mom it sounded terrific.

The waitress came back with our food and pulled a bottle of Tabasco out of her apron and handed it to me.

"I don't know how you eat that stuff, Dave," my dad said. "Just looking at it gives me heartburn."

I took a bite out of my bagel sandwich, and then immediately took it apart and doused it with Tabasco.

On the hike, I pulled away from the group to smoke. It was sunny and brutally cold. We were all wearing our winter gear, February clothing for October weather. I watched Anne. She was perfectly nice. A perfectly nice woman. She was giving and sweet and laughed at my dad's jokes. I found myself being aloof with her and felt guilty about it. Pulling away from someone who loved me; I was watching it happen.

Sally loved Anne and came to her as easily as she came to me. We scrambled down a hill of limestone rocks to watch Sally swim after sticks in the small creek leaking out of the dam, her dog cheeks chuffing as she swam.

Mom came up to me and whispered, "We really like her. She's good for you."

"I know," I said.

We skipped the movie and ate at a chain Italian joint near the intersection of 270 and 315. The place had a vast open kitchen, and so we watched the youths working there spinning dough in the air and the blackening chickens rotating in the giant open ovens. Clouds had slid in, and the day had turned gray. We ate plates of spinach salad, gooey artichoke dip, and pasta, chasing it with wine. My dad told a story of the first day at the restaurant he used to own and the first time someone had ordered wine there.

"This was small-town Ohio, you understand. I wanted to make a big deal of it. Show the people some class. So I took a napkin and draped it over my forearm and cradled the bottle and made a show of displaying the label, then unfolding the little knife from the wine opener to pare away the foil. And I took that knife, and as I was peeling the foil away, the knife slipped and went straight into my hand, right between the thumb and index finger. This was our first bottle of wine at our brand-new restaurant. I bit my lip, wrapped the napkin around my hand, and poured the wine into everybody's glasses, placed the wine in the center of the table, went into the kitchen, and cursed like hell." My dad's eyes glinted as he told the story. "One of the waitresses came up to me, shocked, and said, 'I didn't know you knew those kind of words, Mal.'"

We all laughed. Then I realized that I didn't know what my dad did for a living currently. I panicked, and my mind telescoped away from the moment. I became nostalgic for the moment happening right in front of me. If my amnesia hadn't been because of the Lariam, if there was something biologically wrong with me that the doctors couldn't find yet, and whatever it was happened again, then this moment would be gone, a neural chain of memory scattered through my brain, like a jar of marbles spilled out into a ballroom. I'd have to begin the long process of collecting each marble and making sure it was in its right place inside the jar. If I lost my memory again, someone would have to tell me the story of Dad telling the story about the wine.

Each moment of happiness was now prey to this melancholy. What use was a specific moment of happiness if it couldn't be recalled, exhumed from the gray matter of the brain to relive the happiness? I excused myself and went outside for a cigarette.

That night Anne and I undressed and climbed into bed. She was wearing white cotton underwear and a white cotton bra, and I took them both off her, sliding the panties down with my toes into the cold foot of the bed. We kissed, and she wanted to know if this was going to be okay. I wanted to screw her. I wanted to screw her to prove that I was okay, that I could handle it, that I was still worth screwing. I wanted to see if sex with her made me remember having had sex with her. It was awkward, and I didn't know if it was okay for me to laugh at the awkwardness. I told her I had a condom. She reminded me that she was on the pill and we'd stopped using condoms ages ago. We rolled around, and it was nice. Not cathartic or exhausting, but more of the same kind of pleasant. She fell asleep with her cheek against my chest, which kept me from getting up to get a cigarette.

I lay there awake, wary that the darkness would begin to mutate. My anxieties kept my brain alert and spinning. Anne had just slept with me, but I wasn't the same me from before, so she had just cheated on me, with me. How many of us were there in the bed? The me she thought she knew, the me who I was, the me who I was afraid of being, and her, both the her she thought she was and the her I saw her as right now (and maybe even the her she thought that I thought she was). That's at least six of us. I'm surprised the bed could stand it.

The next day I took her to the airport and had coffee with her at the Chili's after she checked in. I felt terrible. I didn't want her to go. If she stayed with me, I'd know some of the outlines of who I was as an adult. We chatted about my plans, about her plans, about India. She produced a book from her backpack.

"It's for Veda," she said. "You said he needed this for his dissertation, so I stole it from the library."

I didn't remember asking her for it. It was a poetry anthology from the 1950s, all British modernist poets. The pages were yellowing, and the cover was the kind that libraries use when the original one has gotten too worn. Anne pointed to where the magnetic strip had been ripped out of the book. She was sneaky.

I thanked her, and she checked her watch, shouldered her bag, and hugged me.

"You're going to be all right, baby," she whispered in my ear. "We'll get through this thing."

I stood by the arcade, where kids stood in front of bleeping shuddering machines, and watched her go through security.

A week later I wrote Anne an e-mail. I told her that I was going through a tough time and that I didn't think I could be what she needed in a boyfriend. The truth was that she had seen me weeping and trembling, seen me terrified to move off a scrap of lit-up carpet, seen me at my worst, seen me at my weakest, and I couldn't forgive her for it. Or that was a part of the truth. The more I found out about myself, the more frustrated I was with who I turned out to be. I had tried to be a good boyfriend to her that weekend, but it never felt more than a role in which I'd been poorly cast. It just felt so obvious I was faking it. If I was going to be someone better than I had been, then I needed to be honest. While I didn't necessarily know what the exact parameters of my truth were, I did know that being with her hadn't felt like honesty.

I didn't say any of this in the e-mail. I just broke up with her and apologized for having failed her.

For most people, Lariam does exactly what it's supposed to do: it concentrates in the bloodstream and stops the malaria parasite from developing. But for some, through a hiccup of neurochemistry, Lariam pools in the brain. When this happens, usually the worst side effect is a string of extremely vivid dreams and perhaps some mild confusion.

What Lariam is very good at is crossing the blood-brain barrier. The blood-brain barrier is an incredibly effective, tightly knit fence of capillaries that keeps the majority of the world's evils at bay. There are, as a proportion of the world's diseases, really very few that can infect the brain directly (syphilis, meningitis, and multiple sclerosis being three of the most notable), and when the brain does get infected, it's just as difficult to get medicine through to treat it. Very few chemicals can slip past this barrier, but most of those that do, we know well: nicotine, caffeine, and alcohol.

When Lariam pools in the brain, it begins to act as a neurotoxin. It can interfere with synaptic activity by interrupting two very specific kinds of protein gap junctions. These protein gap junctions are small channels created between the cytoplasm of two adjacent cells, enabling superfast communication. Scientists type these junctions by their size, and Lariam affects two very specifically sized ones. One is found in the areas that process information from the eyes, and the other is in the vestibular system, the system that processes all the data from your senses and establishes your balance and body's response to them. Lariam can nestle into these protein gap junctions and scatter the data that passes through them, like putting your thumb over a hose's spray.

This is just a hypothesis. It's hard to know what Lariam is doing as it accumulates in the brain, because in order to measure and study it, the patient must be dead. What is being actively studied is the many possible mechanisms in the brain by which Lariam can cause damage. The weird upshot of this is that scientists are learning more about how the brain functions by analyzing how Lariam damages it.

When a person is susceptible to Lariam's worst side effects, he sees things that aren't there, and his body's reactions are skewed at the same time, like being in a house of mirrors while taking LSD. Input and output are affected. You can't be sure of what you're seeing or how your body will react. Receptors are clogged and the brain is poisoned.

It was my last week in Ohio. In preparation for my return to
India, I pulled a copy of the project I had proposed when I applied for the Fulbright grant. It was lengthy, and I tucked it into a folder and walked the mile and a half to the small strip of downtown businesses. I parked myself in a coffee shop. Rutherford B. Hayes, the president, was born in my hometown. His childhood home had been torn down in the 1980s, and in its place was a BP station. In the front was a small rock garden with some anemic trees surrounding a plaque that designated the historical significance of the location. The coffee shop was right next door to the plaque.

I got a coffee and flapped the file folder open. I read the letters of recommendation first. I was praised as being competent, quick-witted, genial, and a perfect fit for the kind of research I was going to do. I read my proposal and was quickly confused. It was full of jargon that I skimmed over. I had been working on a novel that had many scenes set in India, and so I wanted to do a series of interviews to establish the spoken grammar that Indians used when speaking in English. I would accomplish this by doing grammatical analyses of the other languages the speakers spoke and see how those grammatical structures influenced the way they constructed their sentences in English.

None of it made sense to me and now was doubly daunting. I was getting an outline of some things that I for sure didn't know any longer, and I hadn't known until I opened that file that I hadn't known those things. There were phrases like "dental plosives" and "nasalized vowel sounds." This application was like the list of intimate details from my ex-girlfriend Ariel's letter: words that I couldn't find places for in my brain. This task of reclamation was so much more difficult now. I went outside to smoke a cigarette.

While I was outside I noticed that across the street, men in white pants and white short-sleeved button-down shirts were going inside a bar. It was nine in the morning.

I went back inside, packed my stuff, and crossed the street to do the same. It was a narrow but long place called the Backstretch. Tin signs advertising beer-branded race car teams hung on the walls. There was a pinball game in the back. Ashtrays dotted the bar, so I pulled up a stool and lit up a cigarette, glad to be smoking and not shivering.

The bartender came over, and I ordered a beer and a scotch. I flapped the file folder open again and read my statement of purpose. When the woman put the drinks in front of me, she asked me if I was Dave MacLean.

It must have seemed like a simple question to her.

I looked at her.

"It's Debbie. Cat's sister," she exclaimed, and then came around the bar and wrapped me up in a hug.

I told her that I didn't know she worked there, which was the truest thing I could've said, and then she went into a long story about the chain of events that had led her there. She had two babies with Mark, and they traded off shifts at the bar and were thinking of buying the place from the old woman who owned it. She wanted to know all about me, what I'd been up to, what I was doing back in town.

I told her that I was back for just a bit before returning to India on a grant.

"India. Wow. That must be really intense."

"You have no idea."

She lit up a cigarette. "I'm supposed to be quitting, but it's so hard working here and everyone is smoking."

"Who are these guys in the white uniforms?" I asked.

"They're Honda guys. They work the late shift and come here right after." She stubbed her cigarette out and pushed the ashtray away. "You worked there, right? That year you took off between high school and college?"

I nodded in what I hoped was a convincing fashion.

"Between these guys and the college students, this place can make some real money."

"Could I get some quarters for the pinball machine?" I asked.

She picked up the cigarettes and almost had one out of the pack before she caught herself and put it down. "It's broken. I need to call that guy."

She went to the other end of the bar and picked up a cordless, punching in numbers off a card taped to the bar.

I went back to reading about myself. I had been very confident that the opportunity of the grant would allow me the time and resources to flesh out a novel in a way that wouldn't perpetuate in American popular culture the stereotypical depiction of the Indian accent. I finished my drinks and left two singles on the bar. Before I left, I slid the lone quarter I had into the pinball machine. It spun through the dead machine and clunked in the empty coin box.

I ended up having a car for the rest of the week. Dad was able to carpool with Mom, and my range was extended considerably. The Gold's Gym on 23 was having a special where you could get a two-week trial membership. I went in and signed up, figuring I'd work out every one of my last days in Ohio. I filled out all the forms and was given the tour. At noon on a weekday, it was packed with retirees. I was the only person there under sixty. After pulling on my shorts and shoes, I made a cursory attempt at lifting weights, then started up one of the treadmills. Almost immediately my body fell into rhythm. The weights had felt alien to me, but running felt preternaturally easy. I plugged in my headphones and listened to *My Aim Is True* by Elvis Costello. I kept pushing the machine to go faster and faster. Time slid by, as did my anxieties. I kept punching the buttons, making the grade higher and higher. My legs moved easily. They were strong and able and happy with the *thud thud thud* as my feet struck the black conveyor belt over and over. By the time I finished, I was drenched through with sweat and my lungs burned, but I was maniacally happy. I ran for a little over an hour that day and did eight miles. I blasted the Elvis Costello album on the way home, the windows in Dad's car down as I smoked.

The next day I woke up, and it was like someone had driven icicles into each of my shins. I hobbled downstairs and stretched them out, throwing my workout clothes in a plastic bag. I ran again for eight miles, then went to the Backstretch and had drinks and cigarettes with Debbie.

Running, smoking, and drinking. The rest of the week melted into this pattern. Here were things I could control. I was addicted to them. I could exhaust myself by punching the appropriate buttons on the machine, and I could mediate the chemicals that came into my body through the cigarettes and scotch.

The Friday before I left, my dad and I went to a matinee of some terrible action flick, and then went to a bar and drank scotch. I was limping a little, and my shins were killing me. Dad told me that I didn't have to go if I didn't feel ready. I told him that staying in Ohio was not an option. I needed to get back to my work, I told him. I didn't understand what my work was, but I hoped that if I started doing it, I'd fall into a groove.

"When did I get so interested in India?" I asked.

"College. You first went with two of your buddies after you graduated. They were there for two months, and you stayed an extra two months, bumming around."

"Was I on a program or something?"

Dad signaled to the bartender and got us two more single malts. He liked to call them "drams" when ordering in public. He crunched the ice of his empty drink while he waited for the refills. "You were just out there. You'd call every once in a while and tell us you were okay, but we didn't know what you were doing other than that. You read a lot. Even that James Joyce book that no one can read. You read it in three days on a train trip from the north to the south."

"Were you worried?"

"Not really. You always took care of yourself. This was before everyone had e-mail, too, and so we were used to being out of contact more often."

The bartender came over with the bottle, filled our drinks, and asked us which one of us was driving. My dad signaled that he was, and the guy tipped the bottle up for an extra splash in my glass.

"Why India?" I asked. "Was I into religion or yoga, or something like that?"

"No. Not that stuff." Dad took a sip out of his drink and laughed. "You said you liked it because it was the farthest place away from Ohio."

On Sunday, I walked along the dam with Mom and Sally. Sally bounded up and down the earthen mounds shored up to contain floodwaters. Mom didn't want me to go back to India unless I was positive that I was all right.

"I knew something was wrong right when you started taking that drug. You were acting so weird," she said, her eyes wet. "I hate that I didn't say anything."

"I was probably stressed out. I don't think the drug was affecting me that early."

"You were just acting so strangely. I should've seen it. I should've said something."

"This is when I threw the book?"

"You've always been high-strung, but this was different." The wind was bitterly cold, and it came up off the lake, icy and mean. There were tears in our eyes as we blinked through the wind. She sat down and stared at the sluggish water. I hunched down behind her, and even then it took me three tries to successfully light my cigarette.

She grabbed the pack out of my hand. "Camels. These are what my dad smoked." And she tapped one out of the pack and smoked with me.

My dad contacted some lawyers about suing Hoffmann–La Roche, the maker of Lariam. The lawyers told him that for years people had been trying to put together a class action suit, but the side effects were too varied to be consistently linked to the drug. Since most people were prescribed the drug before traveling, the symptoms would evince so far away from the prescribing doctor that causal connections were difficult to establish. There have been only a few settlements regarding Lariam, including an Ohio man who came home from a safari acting strangely. He went down to the basement for a gallon of milk and instead grabbed a shotgun and killed himself. Roche settled with his spouse for an undisclosed sum.

Lariam poisoning often sneaks by doctors and gets diagnosed as something else, such as schizophrenia, which usually manifests itself in the sufferer's early to midtwenties and has some of the same symptoms (hallucinations, disassociative behavior, suicide). The Peace Corps, ground troops of the military, and travel abroad programs are peopled nearly exclusively with the young, and cases of odd behavior often get attributed to young people reacting to stressful situations.

My parents and I sat at the Chili's in the Columbus airport sipping coffee. I had checked my bags and now clutched my paper ticket. My mom had her hand on the back of my neck, and then she was tucking my hair back behind my ears and pushing her thumb on the crease between my eyebrows that formed when I was nervous. We traded passing bits of minor importance: Ohio politics, basketball games, my dad's new dietary restrictions. I had a layover in New York for seven hours, so Betsy was going to pick me up at the airport and make me dinner. Mom wrote Betsy's phone number and address on a scrap of paper she tore from her address book. Before she handed it to me, she wrote her own address and phone number on it as well.

"Keep this in your wallet," she said.

"I should get it tattooed on the bottom of my foot," I joked.

Mom excused herself and went to the bathroom.

While she was gone, Dad asked if I wanted to split a plate of nachos.

"When she comes back, say that you ordered them," he advised.

Mom came back and gave a disapproving glance at the appetizer in which Dad and I were wrist deep. Sometimes, you can actually see, like a neon scar across her forehead, my mom picking her battles. She let this one slide and folded her hands neatly in her lap. When the waitress came by, my mom said she'd enjoy a refill on her water.

My mom was fiddling with her black butterfly sunglasses, opening and closing them, examining the tiny hinges.

"Those are mine," I told her.

"They most certainly are not."

I wiped the cheese grease off my hands on a napkin and reached for them. She pulled them out of my reach. "C'mon," I said.

"David, these are mine."

"I can't believe you're being like this. I'm going to India, and you're not letting me have my sunglasses." I exhaled noisily and

said, "Classy."

My mom, her cheeks flushed, leaned forward. "I have the receipt for these up in the car. Do you want me to get it for you?"

"Whatever. Keep them. It's not worth it." I busied myself zipping up my backpack and checking the straps. "I should get going."

"We have time," Mom said. "I'll get the receipt and show you. I'm not stealing from you."

"It's fine. They're yours. Whatever." I stood up and gestured to the desiccated remains of the nacho plate. "You want me to pay for those?"

Dad pushed back from the table. "We'll get it."

"Okay. I'm going." I leaned down and hugged Mom and then Dad. "I love you. Thanks for everything." My voice was pulled down by the argument, and so everything sounded sarcastic.

"If you want them, take them," Mom said. "Don't leave like this."

"It's all fine. I'm fine. I love you both."

Brittle and argumentative, I was furious with them. How in the world could they allow me to leave? Should I have thrown a book across the Chili's to get their attention? My parents needed to stop me. I wanted to be back in a hospital, strapped down, with nurses who fed me and thought everything I said was funny. I was safe there. I wanted to be safe. I wanted to be surrounded by people who acknowledged the very simple and obvious fact that they knew what was better for me than I did.

As soon as Lariam was made available, reports of extreme behavior started popping up. In 1992, a Canadian soldier in Somalia savagely beat a civilian to death and then attempted suicide in a holding cell, resulting in permanent brain damage. The Canadian soldiers in his unit referred to the day they were supposed to take their weekly dose of Lariam as "Psycho Tuesdays."

The first time I went to India was in 1998, after my graduation from college. I traveled with two buddies, Duncan and Emil, visiting shrines for a couple of months, and then traveled by myself for a couple of months. I was prescribed Lariam by the doctor I saw at Ohio State University's Travel Clinic. She told me to take it every Thursday. She called it the *Seinfeld* drug since that was the day NBC broadcast the show.

I was three months into my trip. Duncan and Emil had gone back to the States a month before. I was in the southernmost state of India — Kerala — and I was coming out of a dance recital when a man shoved a flyer in my hands. It was for an ecotourism lodge a bus ride away. He and his mom ran the place. I was heading that way anyway, so I told him I would get a room for the next week.

It took a bus and a mile walk, but I got to the lodge and found it was a house — their house. I was greeted by the mother, who'd been told to expect me. She gave me a tour of the grounds. It was a speculative tour: Here was where the second house was going to go. There was where the garden was going to go, which would provide all the vegetables for the meals. Right here, she told me, she thought she'd have a row of hammocks for people to take their rest.

I had been reading *The Catcher in the Rye*. While she gave me the tour and I nodded politely and gave my opinion as to what exactly we Western Tourists wanted, there was a voice in the back of my head that sounded a lot like Holden Caulfield's. It was calling me a phony.

The woman lent me a bicycle, and I rode on the flat roads for miles, trying to exhaust myself. My thoughts were beginning to spin. I rode back to the house, and without changing clothes, I jogged through the large rubber tree forest. The giant trees spread out in a green canopy, each with a tube hammered in its side and a bucket collecting the slow, steady drips. I stopped, folded over, gasping for breath — and I felt something in the forest, something with teeth, that was looking for me. I sprinted back to the house.

When I got back, I was afraid of the book. I took the copy of *The Catcher in the Rye* and shoved it between the cushions of the living room sofa, and I locked my door to keep it away from me.

I knew where the knives were. She'd shown me the kitchen. We were alone. Her son was still in the city up north, handing out flyers. I could kill her. I could get away with it. I'd do it if I wasn't such a phony. I could see her dead already in my mind. It was like I had already done it, was already guilty of it, and all I needed to do was finish it. Go to the kitchen, get the knife, and stab the old woman. It had already happened. All I needed to do now was start it.

I took my ballpoint pen out. I shoved the tip of it underneath my thumbnail. It left a mark a quarter of an inch long under there. The pain was excruciating, but it centered me. For the rest of the night, as soon as I felt my thoughts spinning out of control, I pushed down on my thumbnail until tears came to my eyes.

I made it through the night like this. The next morning I asked the old woman for a Band-Aid, and then I checked out.

I had never had homicidal thoughts in my life before that night. I told myself it was because I had been alone so long.

Solitude was what was driving me. I figured the Lariam had something to do with it, but only because it gave me such crazy dreams. I blamed myself; I felt terrible; the images in my head of her bloody corpse were too vivid. I was ashamed that my brain could produce such things.

From 2002 to 2004, two UPI reporters, Mark Benjamin and Dan Olmsted, filed over forty articles about Lariam and its effects. The reports describe the men and women who served in the US military in malarial regions, such as Iraq, Rwanda, Liberia, and Afghanistan, and then came home and killed themselves and sometimes their family members as well. One soldier in Colorado held his wife at gunpoint in their backyard and told her she had to watch as he shot himself in the head. Police responded to the scene, and when they heard a gunshot, they fired, striking the man after he had already shot himself. Their bullets hit him as he was falling.

In the summer of 2002, three Special Forces soldiers murdered their wives at Fort Bragg, North Carolina. All three had just returned from Afghanistan, and all of them committed suicide soon after the murders. In December of that year, another soldier who had returned from Afghanistan came at his wife "like a linebacker," pinned her against a tree, and tried to strangle her. The woman's seven-year-old daughter tried to protect her mother with a kitchen knife.

Three other Special Forces soldiers who had served in Afghanistan killed themselves that year.

The Department of Defense (DOD) denied for years that Lariam had anything to do with these deaths. The DOD, when confronted with questions, pointed to Lariam's FDA approval and that the Centers for Disease Control hadn't withdrawn its support of the drug.

Olmsted and Benjamin dug through thousands of Roche documents related to the drug and found that in Roche's Safety Report for 1994, the company did address the suicides that were being attributed to the Lariam. The report said that while Lariam could cause depression and that depression could involve suicidal ideation, the suicides themselves were more likely the result of "the progressive breakdown of traditional values" and not of Lariam use.

159

PART FOUR

I can't even hardly remember what happened. It's like a gap.
But it left me alone in a way that I haven't gotten over. And
right now, I'm afraid. Afraid of walking away again.

— *Travis Henderson*, Paris, Texas

I drank scotch throughout the three flights it took to get to Mumbai, the little empty bottles of Dewar's filling the seat pocket in front of me. I studied my brain. I wanted to catch it slipping. I wanted proof that I had no business going back to India, proof that would permit me to run away from the next connecting flight. I wanted to be thrown back into the arms of the people I'd left in Ohio. I was practically begging the flight attendants to intervene, or at the very least cut me off. I ran out of money before they got their chance.

We touched down in Mumbai at one a.m. They cracked the door, and the cabin was filled with humid air tinged with diesel exhaust.

I filed off the plane onto the tarmac and followed the line of travelers into the airport to claim our luggage and attend to the demands of the customs agents. Sleep deprived, jet-lagged, and a little drunk, I queued up in the long line of exhausted, sweaty, wide-eyed white people. The line for Indians moved much quicker. Each one of the foreigners presented themselves to one of a long line of belligerent agents. I stepped up when it was my turn and molded my face into a mask of unthreatening affability. The man, in his white shirt, barely looked at me as I fed him my passport and declaration forms.

"You've come back?" he asked.

I nodded, then realized that he wasn't looking at me. "I had a medical thing, but I'm all better now." Why was I telling him this? Was I asking for him to not allow me to enter the country? "Everything's fine," I continued rambling. "I'm all checked out."

"Profession?"

Now my flop sweat was added to the sweat of standing in that humid, hot air. "I'm a writer."

"A writer," he repeated. He seemed to be testing the word.

The man peered at me over the top of his glasses. "But you're on a student visa."

I stumbled for a second, ready to feel a hand clamp on my shoulder and lead me away to some small cell deep in the airport. "I'm a student studying writing. I'm on a Fulbright grant." I was saying everything like a question, the ends of my sentences rising up and flaking away.

The man unfolded my passport and smashed it flat on his desk. He typed into his computer and studied the passport for several seconds like he'd never before seen such a blatant forgery. He grabbed a large silver contraption and stamped my passport in three different places. He didn't say anything when he handed it back to me. I lingered there for a few moments, waiting for some sort of final word of officialese from him.

He looked up and was surprised to see me still hovering above him. He waved his hand at me, as if he were clearing a fly from the top of his drink.

The flight from Mumbai to Hyderabad was bumpy, and the
man sitting next to me was unnaturally calm through it. He
was young, and the heavy black frames of his glasses were
clearly expensive. He had a manila envelope open on his tray
and a scientific calculator that he punched in little flurries as he
paged through documents. The plane would bump and shake,
his calculator would clatter to the ground, but he was unaffected
by anything that was going on. I suspected I knew something
about physics that he did not. We were heavy and air was light;
air was nothing and we were something. I braced myself for the
plunge when the world finally figured it out.

I had stopped at an ATM in the airport and so had rupees
at the ready. I ordered two scotches from the attendant and
dumped them into the little plastic cup. My drinking strategy
now wasn't an attempt to prove I was unsuitable for adult decisions. I was just trying to make it through the flight without
crying. I gripped one of the arms of my seat with my sweaty
palm and grabbed big swallows of the scotch between bumps. I
had Gandhi's autobiography with me, a present from Betsy, but
I suddenly felt it would be pretentious to be the only white person on a flight full of Indians to haul it out. And Gandhi was
dead, and anything associated with death felt like a no-no just
then.

I was flipping through the in-flight magazine when all of a
sudden I realized that I had no idea where I lived. I assumed
that I'd take a cab, but I did not know what to tell the driver. My
chest tightened. With each breath I couldn't take in more than
a teaspoon of air. My thoughts spun. I'd have to spend the night
in a hotel, but even the next day in broad daylight I wouldn't
know where to begin. I knew I lived after a series of flyovers.
Overpasses were called flyovers in India. I knew that. I knew
that to get to my apartment you had to go over three of them.

How many flyovers could there be in Hyderabad? It'd narrow my search, at least. I pulled out my wallet and flipped through it, hoping to find a scrap of paper with my address.

"Are you all right?" the man in the expensive eyeglasses asked. He was shorter than I and immaculately shaved. It was four a.m., and he looked absurdly fresh.

"I can't remember where I live," I said to him.

"You live in Hyderabad?"

"I do. I live in Tarnaka." Talking made me remember the name of the section of town I lived in. "But I can't remember the address. Could I just say Tarnaka to the cab driver?"

"You live in Secunderabad," he corrected.

"No. I live in Hyderabad. Tarnaka, Hyderabad."

"Tarnaka is located in Secunderabad. I have lived there all of my life."

"Okay. I definitely live in Tarnaka."

"I have a car. Do you want me to take you?"

"Maybe. I think I'd recognize it if I saw it. I know there are three flyovers in order to get there. Does that help?"

"Not really. But Tarnaka is not very big. We could drive around."

"I don't want to inconvenience you. It's so late as it is."

"I'm going to be awake anyway." The man smiled at me. "If you don't mind me asking, what is your occupation?"

"I'm a writer. A student. A student studying writing. I'm on a grant."

"Very good. My name is—" He said his name, but he said it very quickly, and it slid out of my mind as if it were greased.

"I'm David."

We chatted amiably. I worried about getting in a car with a strange man. Paranoia started plucking strings in my brain. Who else but a kidnapper would look so fresh at four in the morning? Who other than a murderer was this nice to somebody on a plane?

"I'm going to take a cab." I fluttered my hands in front of

me in a gesture I hoped meant that there was nothing to worry about. "I don't want to be a burden."

"The taximan might not speak English." He brought up his fingers like he was ready to begin a list of reasons that I should accept his offer.

I interrupted him and said as firmly as I could, "I'll be fine. Thank you."

It took another thirty minutes to land. We sat in silence the rest of the time. I stared at the lights of Hyderabad as we circled. Somewhere amid all those pinpricks was a room with my stuff, a place where I might belong.

When I deplaned and we filed across the tarmac to the terminal, the panic built in me again. The man who had been seated beside me had shouldered his bag and was walking briskly in front of me. I kept myself from running after him and begging him to drive me around town. We entered the terminal. The entire sprawling train of us made our way through the empty airport to the baggage carousel.

We left the secure area. The metal detectors and X-ray machines were unmanned, and there was a lone security guard eyeing us as we passed into the main hall. There was a length of metal fencing against which loved ones were piled up, sorting the newcomers with their eyes. A very dark-skinned man's face lit up when I walked by. He had small gold-rimmed glasses and an amazing pompadour of black hair that split into two wings at his widow's peak. He grabbed my arm as I passed.

"The hero returns," he crowed.

I had learned to recognize people recognizing me in central Ohio. This man knew me.

He met me at the end of the metal fence.

"So you have a bag or something? I have a cab outside waiting."

I didn't recognize this person, but I knew his voice. As soon as I heard him, I knew I knew him. It was Veda. I knew Veda. He'd been my Fulbright-appointed facilitator when I came to Hyderabad. He'd helped me find my apartment. The gears in my brain were rusty, but they were slowly grinding to life.

"How did you know to come get me?" I asked.

"You sent me your information by e-mail. You don't remember this?"

"I'm sorry; I don't."

"You are such a kidder," he said, throwing his head back and laughing. "Come on, let's collect your baggage."

In a rusty van that rattled and jumped with every pothole, I watched the city slide by. Everything was foreign; I recognized nothing. There were palm trees and people sleeping on sidewalks and four flyovers. We made turn after turn, passed a small Anglican church on a roundabout, and passed a marvelously lit-up temple, which was so white it looked wet like fresh paint. The tip of that temple was gold.

Veda was chatting with me, asking about my parents, about Anne. He knew so much about me. My brain spun, trying to salvage memories about our friendship. He told me that his students, especially the females, were waiting for my darshan. He explained that a darshan was when a god revealed himself to ordinary people.

"The girls are really quite taken with you, Hero. You are quite a lady-killer."

"I have something for you." I spun my suitcase around and unzipped it. I came out with the anthology. "Anne got it for you."

He held it in his hands and flipped to the copyright page. "Perfect."

"She stole it from a library."

Veda closed the book. He handed it back to me. "Please have her return it."

"It's already done. Take it. You're picking me up at four in the morning? Take the book."

He wasn't happy, but he flipped to the index and read through the poets. "Please tell Anne that I wish she would not have gone to so much trouble, but that I appreciate it very much. She is a prize. You are very lucky, my friend."

At a random corner, we turned off the main road and passed an array of apartment buildings. We pulled up to one, and Veda said that we had arrived. I had been in the apartment only a month prior, packing my things with my parents, and supposedly I was well at that moment (drugged up and fresh from Woodlands but post-Lariam, at least). And this building still didn't look familiar. It was flat and lifeless.

A gang of puppies wrestled on the street. When we stepped out of the van, they ran at us, squirming and nipping at our hands and pant legs. I pulled at their ears and rubbed their scabby heads.

Veda paid the driver, then clapped his hands at the dogs. They fled, whimpering.

"Filthy beasts. Full of vermin."

Veda banged on the gate. After a while, a thin man in a tank top and a length of cloth wrapped around his waist climbed off a cot, unlocked the gate, and swung it open for us.

We took the elevator to the top floor and then went up a narrow flight of stairs. This was more familiar. I'd climbed this stairway before. We stood at the door to a flat on the roof, and Veda handed me a set of keys and a plastic bag.

"There's toothpaste and some milk and noodles in there. Enough to last you for the day."

I thanked him.

"Okay. I must be off. Perhaps get some sleep before the monsters wake up and start causing terror."

"You have kids?" I asked, embarrassed that I didn't know.

"I will ring you later on your mobile, and we can have dinner tonight at your favorite restaurant."

I grabbed Veda, and I learned that he didn't like to be hugged. After unpeeling himself from me, he waved and disappeared down the stairs.

Immediately he came back up.

"Hero, there is the topic of the rent that we must cover. I paid it for you while you were gone, and . . ." He dodged his head back and forth.

"I'll pay you back. Let me go to the bank tomorrow, and I'll pay you back. Of course."

"No problem. Pay me whenever."

"I'll pay you tomorrow," I said.

He disappeared again, and I realized that I didn't ask him how much my rent was.

Alone, I lit a cigarette. I stayed on the roof surrounding my apartment because I didn't want to smoke inside. The morning was bruising the eastern edge of the sky. Pacing the roof in the darkness, I watched my brain. It was wired and spinning from the flight and from Veda.

I had a favorite restaurant.

I had rent that needed to be paid.

What I wanted was to walk into that apartment and find all of my stuff and be able to click into my life like a LEGO block. I had books in there with my own handwriting in the margins that I didn't remember reading. I'd have to read them all again. Make new notes. But was that really unusual? Who remembered every part of every book he read? But I needed to remember. Every unrecognized bit of marginalia was an indictment.

I had to reconstruct this life. But what if it happened again? What if I lost my memory again, and then I'd have to do this work all over, read these books again, remember the awful crap from high school again, invest myself in strange pictures again, recollect again everything — my family, my friends, the dog.

What if it hadn't been the Lariam? What if my brain was just wired wrong, and I'd spend my life waking up to this life again and again and again? This hollowness. This need for placement. This need for a restaurant full of people who'd all know me and love me and locate me. I hated this stupid flawed brain that forgot that I'd arranged for Veda to pick me up from the airport, that forgot Veda and Anne and everyone who meant anything to me. It was an ingrate, chemically constructed to hurt people who cared for me. I'd rather throw myself off the roof.

Music.

From three different points tucked deep in the darkness of the city, men's voices warbled and cracked. There was the static crush of bad speakers. The voices echoed throughout the sleeping city, calling people to prayer. The day was beginning. Allah was asking through Mohammed through these men through these speakers for all Muslims to come and worship.

I waited, worried that the songs were in my brain, worried that it was happening again. I waited and watched the three different songs swirl in my brain. I waited and was terrified that the songs would cohere into a theme song from my childhood. But they stayed separate. The songs were not about me, and they stayed separate from me. I was okay. For the moment, I was not insane. I wasn't going to jump. I wasn't going to stick the burning cigarette into my arm. I wasn't going to writhe on the gritty roof and piss myself.

I stabbed my cigarette out on the little half wall that surrounded the roof, flicked it, spinning and orange, to the asphalt seven stories below me, and walked into the apartment furnished with the belongings of a stranger.

Hyderabad and Secunderabad are the twin cities of southern India and referred to as a single entity in the same way Minneapolis–St. Paul are, by which I mean, by everyone except for the people who live there. It's an odd city perched on the Deccan Plateau and not frequented much by tourists, no matter what the tourism bureau claims. Its most beautiful vistas are of the rock formations that dot its borders. After erosion washed the soil away, the rocks were left stacked in awkward towering pillars. The city is both Hindu and Muslim, with the Old City predominantly Muslim. The most famous bit of architecture in Hyderabad is the Charminar (the name coming from *char*, which means four, and *minar*, which means minarettes). It's a tall structure, and you can pay 20 rupees to climb the stairs and look down at the traffic snarled around it. In the spring, the fruit sellers are out with long cartloads of Alphonso mangoes, and when they are viewed from above it's like the sides of the street are paved with lumps of ripening gold.

A few blocks away from the Charminar is a tiny shop front; in it there are six men squatting over two six-inch blocks of metal and wood. Sandwiched between the blocks is a sliver of silver. The men beat on the blocks with small hammers, a tiny *ping* with each hit, creating a rhythmless rhythm that you could get swept up into very easily. They fall into patterns with each other and then fall out just as quickly. They beat the silver into wafer-thin sheets that are then placed on fancy desserts.

Instead of doing what my grant had said I was going to do, completing interviews with people and examining grammatical patterns, I went down to this shop front and set up the fancy equipment I'd found in my flat and recorded these guys for hours. I'd let their labor fall into a rhythm in my head and then realize that there was no such pattern, that it was all chaos, and then I'd find the pattern again. Every once in a while one of the guys would hand me a thin square of silver, and I'd set it on my tongue. The wet metal melted, and it was salty. It became my favorite part of Hyderabad.

I was in Hyderabad by crazy chance. When I applied for my Fulbright, I had wanted to be in Benares, a big city for tourists and pilgrims alike.

Benares, also known as Varanasi, has the holy river Ganges flowing through it. It's got the Westerner's checklist of Indian exotica: ubiquitous religious iconography; people bathing in public; a labyrinth of narrow alleys choked with free-roaming garlanded cows; holy men on every corner; monkeys that steal food from your plate; beggars who race you to your hotel before you check in so they can score a kickback; a dozen different scams going on in the streets at once; the gnarled fists of lepers cradling their begging bowls, beseeching passersby; parades with corpses lifted above the crush of people, winding down to the river, where the dead are settled onto biers of logs and set alight. American tourists love Benares for its exotic wildness. Even Mark Twain visited the place.

I remembered being there in 1998 when I traveled after college with my friends Emil and Duncan. I had been sober and overwhelmed by the place, but Emil and Duncan needed more and sought out local hallucinogens. Their eyes were the size of saucers, trying to take it all in visually, pharmacologically.

Benares is a city hyperdense with sense data, and it was where I had wanted to spend a year writing when I applied for the grant in the fall of 2001.

The form asked if I had any university affiliation in India. I didn't, and I didn't have any idea how to get affiliation without cold-calling Indian universities. I did a search on Indian academics and Fulbright and English literature, and I came up with Meenakshi Mukherjee. At the time Mukherjee was teaching at Berkeley. I sent her an e-mail asking if she knew how I could get university affiliation in Benares. She wrote back and apologized that she didn't know anyone in Benares, but that she did know of a literature professor who had done a Fulbright at Stanford and that he'd probably be happy to volunteer to be my advisor.

Mukherjee is a giant among academics. Her books on India's postcolonial literature are bedrock texts, heavily referenced by other academics. Had I known that, I wouldn't have so blithely e-mailed her. But I did, and she graciously helped me. If I had received the Fulbright and gone to Benares and had the amnesia episode there, a tourist hot spot with the train stations crawling with stoned foreigners, and the touts and dacoits who prey on them, surely I would not have been able to find the immediate assistance that I received in Hyderabad. It's because Mukherjee didn't know anyone in Benares that I was safe. Because the person she did know at the University of Hyderabad had a graduate student, that student became my Fulbright-appointed facilitator. After he had helped me set up my life in Hyderabad, I became friends with him. When I was found in the hospital, he rounded up every white face he could find so I wouldn't feel alone. When I came back to India, he picked me up from the Hyderabad airport at four a.m. with a plastic bag with toothpaste, milk, and ramen. As far as I knew, Veda was my only friend in Hyderabad.

I called him the next day and the next, and we began a ritual of conversations and dinners that would last me nearly the rest of the year I spent in Hyderabad. What I knew about Hyderabad, I learned from those dinners. I'd leave my apartment at seven p.m. and meet him at this little place that was dark and covered with mirrors. Walking in, I'd always feel as if my eyes would never adjust enough to see. The host was supremely attentive and escorted me to my table immediately. I'd walk to my table-that-is-just-now-ready-this-way-please nearly blind. I was perpetually ten minutes early. I kept thinking the walk to the restaurant would take longer than it did. I'd watch the clock and count down the minutes until I could leave the apartment and actually have someplace to be. I'd come in and be seated, and before my retinas had dilated enough to take in the mirrored walls and globed candles at each table, I'd ordered a beer and a whisky. I'd have the whisky down before Veda got there.

Veda and I would order baskets of naan and share a dish of palak paneer or malai kofta, and he'd always be surprised that I'd want rice as well. It was part of our ritual for him to point out that rice did the same thing that naan did and that it was wasteful to order both. My part of the ritual was to nod and say that next time we'd only get one of them. I paid for these meals. The bills for them rarely went over 400 rupees, which was about $8, depending on how much I would drink. My grant wasn't for a lot of money — the Fulbright people figured that for the grantees, living on a "local salary" was part of the experience — but Veda was a grad student with two kids, so I figured I had a lot more disposable income than he did.

I also owed him.

Our conversations would slalom through various topics. He had a friend who was obsessed with the literary critic Stanley Fish, and he'd rattle on about Stanley Fish, and I'd pretend to know who Stanley Fish was. Then I'd talk about the books I was reading, and he'd pretend to know who Barry Hannah was. Then he'd tell me about the Russians who were set to attend the school in the spring.

"Some very strong and lovely opportunities for you, Hero," Veda would say and wink.

Veda worked at the Central Institute for English and Foreign Languages (or CIEFL — "sea-full," as everyone called it). It was both a college for locals as well as a place foreigners came for intensive foreign-language study. These foreigners were not coming to learn Hindi or Urdu or even Tamil or Telugu. They came from West Africa for English, they came from Spain for Russian, and they came every year from Russia for a two-week boot camp in conversational English. The Russian groups were always almost entirely composed of women. Veda was deeply unhappy in his marriage, and he wanted to cheat by proxy.

Veda's misery was something he polished and wore like a tie clip. His marriage was awful; his children were poorly behaved; his dissertation was going to be unfinished forever. I never was able to get a handle on how old Veda was. It was one of those things that I was embarrassed to ask because I was sure I already knew but had forgotten. He was probably in his early forties. At least he sat in chairs like a man in his early forties. Veda liked to say I was bright and beautiful and young and could have sex with any and every young thing. Sometimes dinners would devolve into him explaining how great and perfect I was, with me then trying to talk him out of it.

The bill would come. Veda would reach for his wallet. I'd wave him off, and I would pay it. We'd leave the intense dark of the restaurant and enter the purple beginning of night. He'd offer me a ride on his scooter. I'd say that walking was the only exercise I was getting. We'd shake hands; he'd kick-start the scooter and take off. I'd stand there for a few moments, trying to brainstorm things I could do. Rickshaws would accumulate near me. Men poked their heads out of the black and yellow exhaust-belching things and called to me. I waved them off, and for want of anything else to do, I'd go back to my rooftop flat. I'd smoke cigarettes, and with the CD player going, I'd try to memorize Merle Haggard songs. After an hour or two of doing laps smoking and softly singing "Big City" or "I Can't Hold Myself in Line" to the night sky, I'd climb into my cot, down some Valium, and will myself to sleep.

I was still haunted by my hallucinations. I had been denied entrance into some higher order. When my brain malfunctioned, this was the story that got dredged up out of that two pounds of gray matter. Lodged inside of me was a feeling of inadequacy. It was a feeling ingrained as deep as the biochemical level.

I came back to India feeling unworthy, feeling like I needed to put things right. My first job was to thank the people who had helped with my recovery. Veda, I saw almost every night. There was Geeta, who I planned to visit in a week or so, who sent me e-mails every single day, and who may or may not have been flirting with me. I told myself that I was visiting her to reconnect with a fellow Fulbrighter, that my visit wasn't only about being close to a beautiful woman in a bikini. Geeta was a singer, so I told myself it'd be good to spend time with a fellow artist.

But there were also Mr. DeSilva and Sampson, the two men who came to Mrs. Lee's guesthouse and prayed over me. There was Dr. Pat, the American who got me transferred from Apollo Hospital to Woodlands, but my parents had told me that they were donating to her organization, so I didn't feel the need to thank her myself. There was Richard, the poet who looked like Jim Henson and brought me cigarettes.

I didn't want to go to Mrs. Lee's house. I didn't have the heart to tell her that I wasn't a drug addict. It felt like it was more magnanimous to let her believe she had led a successful intervention.

And Rajesh/Josh. I needed to find him. I needed to thank him most of all. If I had ended up on a train that day, if anyone other than Rajesh/Josh had found me . . .

The emotional heft of the hallucinations haunted me, made me feel consigned to this menial existence where I scrubbed my clothes in a bucket on my roof and where my stomach wasn't

good with any of the food. I lost ten pounds within the first
three weeks I was back. The foot pads on my squat toilet were
slippery, and I had already twice stumbled and put my foot in
my own mess. I spent my days reading the books in my library.

I read *The Idiot* in two days. I left my apartment only to write e-
mails to Geeta, coordinating my visit in mid-December; to pick
up milk packets, which I'd boil before pouring over my Wheaties; and to meet Veda for a dinner I'd then spend the rest of the
night squatted over a hole evacuating.

The hollowness was lessening, though. There was so much
clamor in the streets and inside of my intestines that I only felt
the hollowness right before sleep and right when I woke up. I
was remembering more of my life. I was functional, but the
shame remained. In addition to the shame, torpor was taking
over me. Cigarettes, Valium, booze, books, and the hours record-
ing the silversmiths? I needed to get out of my apartment and
interact with the web of people who'd saved me. I needed to
make amends.

I started with Mr. DeSilva. I called to thank him for his help. I was embarrassed and effusive; I hadn't planned out what to say beyond "thank you." I figured the rest would have taken care of itself, that I'd feel some kind of tumbler clicking into place with each expression of gratitude. Instead, I rambled on about how difficult it had been and how lucky I was to have him and Sampson help me. I think he finally invited me to dinner to shut me up.

But then he had a better idea. He asked me to attend church with his family.

So the next Sunday, I put on khakis and a nice shirt and caught a rickshaw to the Anglican church on the roundabout. The service was in English: standing, sitting, singing, saying the things printed in the program in unison. The sibilant hiss that filled the church when everyone got to the "forgive us our trespasses as we forgive those who trespass against us" line in the Lord's Prayer was something I then remembered from my childhood—Sundays upon Sundays at Asbury Methodist. I celebrated each of these little remembrances that came back.

When I declined to go up and take communion, Mr. DeSilva pulled me aside and assured me that the wine and host were all safe, even for foreigners.

After the service, the congregants filed into the parking lot for cookies and punch, and the DeSilvas brought over every white person they could for me to meet and chat with. Mr. DeSilva told me that I was welcome to have lunch with them, and that I would stay with them through the afternoon and attend a night service at a different church before coming back to this church for a celebration. He said I would like the other church even though most of the service would be in Urdu. It was a big time for the other church because we were in the middle of Ramadan. I knew that Ramadan was going on since I lived in a neighborhood with three mosques, but I had no idea why the Muslim holiday would've mattered so much to this other church.

At their house, I was fed mangoes, bananas, and small cucumber sandwiches in the kitchen, where a TV tuned to an evangelical channel babbled on a narrow table in the corner.

The hours crawled by. Slowly the sun descended, and we piled into DeSilva's car. We parked in front of a mosque. There were dozens of men milling around in long white kurtas reaching past their knees and with white knitted kufis perched atop their heads. We crossed the street and walked up a flight of stairs on the outside of a white stone house. At the landing of the second floor we had to squeeze by a giant speaker that was facing toward the mosque. Inside, there was a meeting room with about thirty adults and two young boys in shirts and ties setting up folding chairs on a dusty red painted floor. A chalkboard hung behind a podium with a microphone. The Anglican church had been three times as large, and the priest hadn't needed a microphone.

I asked DeSilva what kind of church this was. He assured me it was nondenominational. I asked again. He said that the woman who founded it was a former Muslim, and she had started up her ministry to let other Muslims see the light. I was in for a treat because she was going to be delivering the sermon tonight. DeSilva let me know that there'd be a translator as well. I wouldn't miss a thing.

The founder of the church came up after we sang a song about Jesus's immense glory and told the story of her conversion. She said that she'd been an observant Muslim who had never missed one of the required five daily prayers. Her husband, who rarely prayed, had claimed that he was going to get into heaven by grabbing her ankles as she was lifted up. Everyone laughed. She talked about what a beautiful religion Islam was, but that it was like a vase: pretty and able to hold the beauty of the world, but shallow and easily broken. She then enumerated the glories of Christ.

I subscribed to the *Deccan Chronicle*, the Hyderabad newspaper, and had been reading about the history of communal violence in India. There had been a new rash of riots resulting from the debated demolition of a mosque in Benares because there was archaeological evidence of a Hindu temple underneath it. Hindus and Muslims were fighting each other in mobs, but in neither of those religions is there an evangelical imperative. Muslims were not trying to convert Hindus, and vice versa.

I'd wandered into something bad. Outside were dozens of Muslims, hungry Muslims, and we were blasting them over the speaker with messages of how flawed their belief system was, a belief system for which they'd spent all day denying themselves sustenance. The service lasted more than two hours, and it was dark when we left. All the bearded men glared at us as we left the apartment building. I tried to make myself invisible.

I was ready to head home. Two services and a lengthy afternoon spent listening to church television had soured me on God, but DeSilva told me that I must attend the celebration the Anglicans were having.

Sampson was going to be there.

Sampson still had a halo about him whenever I thought of him. He had been so kind to me during my hallucinations. DeSilva had been kind as well, but his glow as a living angel had faded during that day of church services. Sampson, though, with that crown of silver hair, was still holy.

So I went to the Anglican celebration.

It was full dark when we arrived. There was a tent set up in the back. We ducked through one of the vinyl sides, and inside was the heady thick smell of meat trays bubbling over Sterno flames. I'd spent the day drinking Pepsi and eating mangoes, so my stomach felt queer, and it was sending tremors of weakness through my arms.

The smell of the food had lost its initial attraction. I went outside and lit up a cigarette, inhaling the hot gravel of the Indian brand deep into my lungs. The grass had been watered earlier in the day, and my feet sank into the ground, the dampness seeping into my running shoes. I tried to make the stars fit into some of the constellations I knew, but I was on the other side of the earth, so nothing made sense. I stood there for a while, hoping that Sampson would materialize beside me, kind of just amble up and point out which star to watch.

I wanted to believe in God the way DeSilva did. I wanted to see someone across the tent, and we'd fall in love instantly. I wanted the straps and the drugs and a nurse who thought I was funny. I wanted something else to take control for a while; it didn't matter if it was Jesus or a lover or a hospital ward.

I pitched my cigarette over the gate into the alley and went back inside to say good-bye to the DeSilvas.

DeSilva didn't try very hard to convince me to stay. He told the people at the table that I was a young student on scholarship studying in Hyderabad and was interested in local languages. Leaning in a little more, I asked DeSilva about Sampson. I'd like to thank him is all. DeSilva craned his neck around.

He'd just left for food. Had I tried the food? It was a shame that I was a vegetarian, because I was missing out on some excellent biryani.

"I'm not a vegetarian," I told him.

"You said you were at the hospital," he said, wiping his hands on a napkin. "We brought you food, some roast lamb in rice. But you said you didn't eat meat."

"I did?" What had I told them? Had I denied the food they had brought for me? "I was a vegetarian in college."

I stood up to leave and knocked into a man holding a paper plate, sending his samosas into his chest and leaving little triangle-shaped grease stains on his polo shirt. I apologized profusely and grabbed a napkin off the table, and only when I offered it to him did I realize it was Sampson.

He eyed me warily.

I checked my desire to hug him. I felt squishy inside. Here was my angel, shorter than I remembered, but still . . . I effused immediately about being glad to finally see him. He pulled back. He asked if I had fully recovered.

"I'm fine now. I sometimes feel a little lost still," I said. "Just sad, but better."

"I would hope so," Sampson replied, and then said good-bye to me and sat down at the far end of DeSilva's table.

My chest was tight. I was worried what I might have said to him when I was hallucinating. Did I tell him how beautiful he was? Was he standoffish because he thought I had a crush on him or because when he first met me I was thrashing in a bed and covered in piss? I hailed a rickshaw. The night had brought a chill with it, and my thin button-down shirt wasn't heavy enough to protect me from the wind that buffeted me in the back of the rickshaw. We navigated the flyovers and the turns with my broken Hindi. Did Sampson know me better than I did? What had he seen?

At the front of my apartment building I argued briefly with the rickshaw man about the rate, then instead of waking the doorman from his tiny cot, I climbed over the gate and walked up the seven flights of stairs to my rooftop flat.

I was treating gratitude as if it were an act of penance. I expected that if I completed the rounds, then I'd have righted the scales, and everything inside my head would immediately feel better. I'd knocked two people off my list, but I still felt miserable, and I spent the night as usual, pacing the roof, chain-smoking, memorizing country songs, and doing my best not to think about throwing myself to the pavement below.

I stood on the platform, the noise and stench and chaos thickly enveloping me. It was early December, and I'd arrived at the station an hour and a half early for my train. Around my waist, under my pants and shirt, I had my passport and money secreted in a pale yellow belt. I had packed the belt two days ago and slept with it under my pillow. That day I had set an alarm so I would have enough time to unpack it, catalog its contents, and repack it again before I left for the train station. I wore the belt cinched so tight that it left marks.

I had my second-class ticket folded in my fist, and it was quickly becoming wet, the blue type smearing into a pulpy mess. I was not hallucinating, I reminded myself over and over again. I knew who I was and where I was going. I had a ticket for a window seat, and Geeta was going to pick me up at the train station in Goa. I was going to be fine. I *was* fine.

I bought an omelet from a man in a stall. The clanging of his pan made me flinch, but I was not hallucinating, and I knew who I was and why I was going to Goa. As long as there was chaos, I was okay. It was when everything congealed into a meaning-making machine producing feelings of unity within the world—if a moment like that came, then I should be worried.

If chaos was comfort, then the Secunderabad train station on a Wednesday morning was like an overstuffed sofa. I choked down the greasy egg and white bread and waited as my train pulled into the station. It was a big monster of a train. The red and blue paint job on the engine had blackened over the years, and in comparison the blue paint of the cars was new and had a heavy lacquer sheen to it. It was tacky to the touch. Above the windows, numbers were painted, and I kept count as the train rumbled by me, peering through the windows, desperate to know if I had a good seat. Everyone told me how important a good seat was—Geeta, Veda, the ticket broker—but no one told me exactly how to pick one. The numbering on train cars varied, and so it was difficult to predict. I'd be traveling through the southern middle of India, coming off the Deccan Plateau and into the lush valleys that emptied into the ocean. I wanted a window seat so that as we traveled, all the palm trees would tilt toward the ocean, and all the tiny rivulets and streams would congeal into rivers that'd join, and then join again, and empty inevitably into the ocean, and it would all push me along, all tilted and rushing, toward Geeta.

I remembered riding trains with Duncan and Emil during our trip in 1998. Second-class cars have six seats on the wide left-hand side and two seats on the narrower right side. I had a window seat, but on the narrow side. There was a group of uniformed soldiers in my car, and they began drinking right away. There was an old man in the seat across from me, and we smiled at each other. We were both alone. I stared out the window like it was top-notch cinema. I was separated from everyone around me in a little glass bubble of impenetrable silence, watching the landscape slide by. I thought of Geeta. I put my bag over my lap and kept staring out the window. The sun kept pace with the train. We were both slowly traveling west, with a billion internal explosions happening each millisecond powering us.

Because smoking was prohibited in the berth, I snuck back toward the open doors and sat with the guys who didn't have tickets. Where those open doors were, with the wind and landscape rushing by, was a kind of lawless zone that the conductors pretended they didn't see. It was like international waters. You could sit on the steps and smoke. The bathrooms were right near where I sat. I was happy to be on the steps and watch the landscape tear by. If I shifted my weight one way, I'd be pitched out into the trees and gravel and die; if I shifted my weight the other way, I'd fall back into the puddle of questionable muck coming from the bathroom. It felt better than sitting in my berth. There on the steps everything was clear for once. I had to maintain perfect balance, and it was exhilarating.

There was yelling coming from the front of the train, and it was getting closer. A wave of sound, and when it crested, my train car went dark. We were entering a tunnel, and people were screaming to hear it bounce off the walls. The train punched its way through the tunnel. Everyone around me was yelling, the wild herd of us taking pleasure in something as simple as echo. In that darkness, I was shouting just to have some part of me bounce off the world.

We pulled into Goa an hour and a half late. I hustled off the train with everyone else, unchaining my bag from the bottom of my seat, cinching my secret wallet tighter, damn near sprinting to get out into the parking lot. A woman who looked like the pictures off my digital camera waved to me from a tiny car rattling with music. As I got closer, the car was a cluster of body parts, arms and heads stacked up in the tiny compact. The woman jumped out of the car and wrapped me up in an embrace. It was Geeta. She was as beautiful as she was in the pictures.

Doubtful that I was going to fit in the little car, I chucked my backpack into the trunk. I opened the rear door, and a petite woman smiled up at me and told me that she was going to have to sit on my lap. I squeezed into the car, she shifted on top of me, I was introduced around, and someone handed me a bottle of warm Kingfisher beer. I recognized the song they were blasting. It was DJ Doll doing a cover of an old Bollywood song, "Kaanta Laga," and it had been ubiquitous in Hyderabad, the controversial video (the young hot singer wore a thong that poked above her tight jeans) playing in every coffee shop, bar, and Internet café. It was all bass and the buzzing high nasal voice of women in Bollywood music.

The woman on my lap was good-looking but a distant second to Geeta, who turned frequently to smile at me. The woman on my lap had her arm around my neck in a way that I wasn't positive was flirtatious. We went to dinner at an open-air place famous for its seafood. Everyone ordered Old Monk and Coke, and I fished out a carton of cigarettes for Geeta. Old Monk is an Indian rum notable for being cheap and dreadful. I drank it quickly to avoid the taste and was rewarded with another one. We had shrimp curry and naan and drunken conversation. The guy on the other side of Geeta was enthralled with her. He was marking out his territory, leaning into her to read from her menu, making sure to clink glasses with her last after every table-wide cheers. His chair was turned a few degrees toward her, so that when he laughed he nearly fell into her lap.

He was tall, thin, good-looking, smartly dressed, and an architect. I was none of those things, but it didn't matter. I was the one who was going home with her.

Geeta had a flat that was the entire second story of a house. She had rooms she didn't even use. She put me in the guest bedroom and warned me to wear slippers when I showered because there was some wiring loose in the electric water heater. Geeta came and lay on my bed, and we talked about my waking up in the train station. I told her that it was terrifying and lonely, and that I felt hollowed out and empty still. I told her that I was having panic attacks and smoking like a fiend.

She leaned in close and tapped the ashes off her cigarette into a soda can in between us. She was in a man's white V-neck shirt and thin cotton pajama bottoms. I usually don't notice clothing, but that outfit is burned in my mind. She told me that when I wrote her that e-mail, she called up the Fulbright people and cried with worry over me. I wondered how long we had known each other before I had gone off the rails—a month at most? Had I meant that much to her in so short a time? I wasn't sure who I had been to have had such an immediate effect on her, but I was fairly certain that getting sick had drawn us closer together. It was as if the worst moment in my life had become a kind of shiny button I could wear to attract women like her. Using personal suffering as an emotional currency normally would be repellent to me, but sitting across from her and her white V-neck T-shirt and thin pajama bottoms, I could see its good points.

When I told her about my hallucinations, I didn't mention Jim Henson. Instead, I described the one where I was an old man in a kitchen filled with my loved ones, my children, my friends, my wife. I lied and told her she was there. I said that it was beautiful and my hallucination was all about universal love.

She pulled back and wrapped her arms around her legs. She told me how worried she'd been about me, about her bad dreams as a result of Lariam and how she flushed her pills when she'd heard what happened to me. I thanked her for writing to me so often. I told her that she'd been a big part of my recovery.

She then said she'd had an affair with a woman who was on the Fulbright with us and that she was confused by the whole thing.

We talked until three, and she asked me to ignore the sounds coming out of her room in the morning; she had to do her vocal exercises every morning. She hugged me and drifted out of the room.

The next day, we borrowed a scooter and took off for the beach. Her landlady had offered her the use of the scooter when she arrived, but Geeta had no idea how to drive one and was afraid to learn. I knew that I had owned a motorcycle when I lived in North Carolina, so I guessed I'd be able to drive the thing. She sat straddled behind me and sang a Magnetic Fields song, "The Luckiest Guy on the Lower East Side." It's a song about an ugly man who gets to drive a beautiful girl around because he's the only one she knows who has a car. She had a lovely voice.

At the beach I bought some cheap sunglasses off a vendor, and we watched a sidewalk tattooist draw designs on a young man with a needle gun powered by a foot-pedaled motor. Geeta's bikini was white, and she was stunning, and even the presence of her white husband didn't keep the men from flocking around her. We treaded water in the surf with a ring of men surrounding us shouting things in Hindi, things that I didn't understand but made her throw her arms around me and kiss me.

"They're calling me a whore," she whispered in my ear.

"Let's go," I told her. "The waves suck anyway."

At her house we made ramen noodles and started having sex. The noodles burnt to the pot and filled the kitchen with smoke. We had sex three times, and she fell asleep. I did not. I was quivering with happiness. In that erotic haze, I felt I had been found. I was convinced that this was the life I was meant to have. Her AC was on the fritz and we were covered in sweat, which the night air chilled. Geeta and I had a connection. It took a lot of awful stuff to happen to us, but here we were—we had *found* each other. I believed that the rest of my life would be spent in that tiny single bed with her. I felt swept up by love, carried by it, and that I'd never be lonely again. I was going to use her to reorient myself; our love would be my true north.

As she slept, I talked to our future children. I told them about how their mother and I first met and how we fell in love and how we made them and how I was a fragile mess of a man until she came along and became something beautiful for me to hold onto and order my life around. I listened to her breathe and mentally cast generations of beautiful children from her— born in a world where all of my genes were recessive.

We stayed inside the next day, the air acrid with the stench of the burnt noodles. We ordered in food and kept having sex. That night as she slept, I talked to our children and grandchildren again, and told them how important love was and how you should always be with someone you didn't think you deserved. I told them that in order to be happy, they had to date out of their league.

The next day, the pale green of her apartment was oddly fun-
gal in the morning light. She woke up and gave me a look that
I couldn't categorize. I hugged her close. She told me that she'd
just gotten out of a relationship with a white guy and that she
didn't think she could be another white man's ethnic girlfriend.
She described how beautiful it was to have sex with the other
woman, who was African American, saying that when two
women of color orgasm together, it's something singular, gorgeous, and an act of protest against the white patriarchy.

She didn't want to have to be somebody's primer on race rela-
tions.

She wasn't ready to date another white guy.

She told me that she needed to brush her teeth, and it'd be
best if I moved my stuff back into the guest bedroom.

My return train ticket wasn't for another two days.

Another Fulbright student was coming into town, and when we picked him up I knew I was obsolete. He was Ivy educated, tall, handsome, strapping, and brown. His name was Kishore. We fit three on the scooter: him in back, Geeta between us, me driving and hanging onto the very front point of the seat with my tailbone. I listened to them flirt. I watched his lame moves and noted that his departure ticket was four days later than mine.

Kishore and I shared the guest bedroom. I hated him as politely as I could.

The next two days I pined for Geeta. I chauffeured her and Kishore around as they flirted. I took them to the beach, and they'd swim and splash each other while I walked on the sand and moped. I was excluded. I thought I'd found something to hold on to, but she turned out to be like Sampson and DeSilva, unwilling or unsuitable to be the handrail that I'd cling to in order to make it through life. We met up with her friends at night for dinner, and Geeta pointed out my sunburn to her friends.

"Who in the world gets pink from the sun?" she cackled, and everyone laughed—everyone except the brittle architect. He'd been moved three seats down from Geeta. Kishore was sharing her menu now. The architect knew what I was going through.

The last day I was there was her birthday. I went to a market and bought yards and yards of flower garlands and strung them all over her sea-green bedroom. She was out studying with her singing guru, and I spent the afternoon balancing on her furniture, hanging garlands everywhere. I finished up just as she was climbing the stairs and putting her key in the lock. I raced into the guest room and opened up a book, trying to look — despite my panting and racing pulse — beatific, sensitive, and forgiving. What I expected for my three hours of work and 300 rupees was for her to realize in a flash the kind of man I was: I was hardworking and thoughtful and willing to overlook her recent behavior in order to make her birthday special. Didn't she realize that she was my true north?

She opened her bedroom, and she realized exactly the kind of man I was: needy, mentally unstable, and unable to take a hint.

I caught a cab for the train station three hours early.

It was a lonely ten-hour ride home, and the entire ride I apologized to our imaginary children, who now would never be born. I thought I'd found something to order my life around. I had wanted Geeta to be what Jesus Christ was to DeSilva. I needed some solid surface off which I could hear my own echo. I kept seeing Geeta's face as she opened up her bedroom, the wave of stink coming off the marigold garlands strung everywhere. The person who had done this big crazy gesture was me. Those other mes—the maniacal country-music DJ or the guy who could never keep a straight face in pictures—were they always acting, always mugging, always hyperbolic because they were there to contain the toxic emotional neediness I had at my center? Geeta had kicked me out of bed, Jim Henson had kicked me out of the astral plane—and I felt like I deserved it all.

I resumed my hermit ways when I got back to Hyderabad. I'd stay in almost all day. I was turning down offers from Veda, from the DeSilvas, to go out. Some days the single social interaction I had was waving to kids on other roofs as they flew kites. I read a lot and wrote nothing but diary entries. Every day I would force myself to leave the apartment and go down to the Internet café, sending e-mails to my parents and sisters, reading CNN and *The Guardian* UK.

At 40 rupees an hour, I inhaled whatever data I could find—movie reviews, sports recaps, anything. I brought take-out Chinese and stacks of magazines back to the rooftop with me.

I needed the world to be stable. I needed people to be simple and trustworthy. I stayed inside and ate Wheaties covered in hot milk for breakfast and takeout or masala-flavor ramen for dinner. I drank Kingfisher beer until I fell asleep, folding my thin pillow over my ears as the elevator engine whined its awful whine.

I washed my own clothes and hung them out to dry, then took pictures of them drying.

After several tries, Veda got me to agree to come out one night for dinner. I got there late and ordered drinks as soon as I sat down. I made a point of ordering rice and naan. Veda handed me a plastic bag.

"I kept forgetting to give this bag to you," he said. "In it are your things from the hospital."

"I already got my stuff from there." I glanced in the bag, but it was too dark in the restaurant for me to see anything.

"This bag is from the first hospital. When they transferred you from Apollo to Woodlands, they gave me this bag with your possessions."

"Thanks," I said.

Veda wanted me to know that the Russians were coming at the end of the month, and he'd be happy to arrange a meeting. I told him that he was using me. I wasn't some sort of cash cow, I told him; I was broke. He thought I was rich because I was white, and that was racism, and I wasn't going to stand for it any longer, I said. I was making a scene, and I was asked by the headwaiter to please be quiet. I ordered another beer and sulked in the dark room covered in mirrors.

Veda paid the check, and we didn't talk for two weeks.

Inside the bag were a couple hundred rupees, some receipts from Mrs. Lee's guesthouse and the Internet place Rajesh had taken me to, a shirt that I thought had been lost, and a pair of white cotton panties. Victoria's Secret, size medium.

The mystery of those missing hours between when I left my apartment and when I was found at the train station kept getting weirder.

One night in December, I caught a rickshaw over to have dinner with Richard, my ersatz Jim Henson who had played stand-in for God in my hallucinations. I hadn't seen him since he'd jammed a stack of poems under my arm as I was being ushered out of the Taj hotel. Richard was in India illegally, having overstayed his visa by three years. He was living with a young woman named Shoba, who was a lawyer and a fiercely brilliant activist for civil rights. She was also oddly deferential to Richard, who seemed less like a lover and more like someone who'd been crashing on the couch for years.

We had dinner out on the roof of Shoba's modest house. She'd been in court all day, but she still busied herself with serving dinner to Richard and me. Richard was pudgier than I remembered, and his beard and his remaining hair were wilder. His conversation was manic. He talked about the Hegel he was reading and wanted to know if I was interested in hearing some of his latest poems.

I asked Shoba if she needed any help in the kitchen.

Our folding chairs made awful noises on the crumbling concrete floor. I sipped at my water, trying not to drink it too quickly since it'd make Shoba get up and refill my glass. When I asked Shoba about her work, she said how helpful Richard had been and how much Richard had read. She wanted to know if she should leave and "let the Americans talk." She was self-conscious about her English. Richard said he'd been working on it with her for years now. He didn't speak any Hindi or Telegu.

When dinner was over and the plates were cleared and Richard and I had both lit cigarettes, the entertainment portion of the night was to begin. It was decided that there should be singing.

Shoba went downstairs and wrangled one of her nieces, a girl of about eleven, to come up and sing for us. The sun was setting, and the bats were darting about in the air. The girl's voice wavered. She moved her feet as she sang. Back and forth. It was an old Tamil film song. Shoba brought up coffee. Richard never thanked her.

We applauded. The niece sang two more songs. Shoba asked Richard to sing. He waved her off, then acquiesced immediately.

"I am going to sing a song by Paul Robeson," he said.

Then he gave us a lecture on Paul Robeson: his life, his activism, the effect of his music on the civil rights movement. Then he told us the history of the song he was going to sing. He took a drag off his cigarette. He sat up straighter, then leaned forward. The sounds of the street below came up, all rickshaw engines and bicycle bells. He began to sing. He pitched his voice very deep and sang "Old Man River." He dragged each note through his baritone.

It was ponderous.

Shoba was dazzled.

When he finished, I got up to leave. Richard was confused; we hadn't talked about his poetry yet. He had all these new poems to show me. I told him I had work to do. We shook hands. I was in a rickshaw on my way back to my flat when I realized I had forgotten to thank him for being there in the hospital when I was ill.

One day in January, I sat at one of the outdoor tables at the café a block from my house. I started making it a habit of going there and having a coffee and a pair of samosas after I had come home from doing my rounds of e-mails at one of the Internet cafés. There was a DVD rental place on the way home, and I could rent bootlegged discs for 30 rupees for three nights. These bootlegs were shot in American cinemas with handheld cameras and then sent to India by way of Taiwan. The American movies would often have Chinese subtitles. The discs were terrible quality. The camera jittered, and the sound was awful. At one point during *Gangs of New York*, Scorsese's epic set in 1863, the guy filming the movie got bored and put the camera down. For fifteen minutes of the movie the camera was pointed at the back of the seat in front of the cameraman. Since I had no TV, these bootlegs, which I played on my laptop, were my main source of current American pop culture — jumpy, barely audible, and subtitled in another language.

I'm not sure what movie I had rented that day, but I was sitting there with my coffee and my samosas when a moped pulled up next to me. The US State Department was releasing alerts nearly every other day warning travelers in Muslim areas to be highly aware; they stressed that I was supposed to vary my routine in order to keep myself protected. It was early 2003. The Iraq war was in its first motions, and North Korea had kicked out the UN weapons inspectors before they could say whether or not North Korea was developing nukes. Paranoia was a side effect of being an American abroad.

The moped pulled up close to my table, and a boy's voice called my name. Three boys I didn't recognize were stacked on the old rattling thing, which was rusty and held together in places with wire. They were street kids with torn clothes. The largest boy, the one driving, was no more than eleven. The one in front of him on the banana seat was five, and the boy sitting in the back was no more than nine. The three of them were very dirty. They'd clearly been joyriding all morning.

"David. How are you, my friend?" The boy in front picked his nose; the one in back smiled at me and waved. "Where have you been?"

"Have we met?" I asked.

"You promised you would come back, and you never have." The middle boy shouted in order to be heard over the traffic.

"I don't think you have the right person," I said. I was wary. Were these boys from that time I couldn't remember? Supposedly I had slept somewhere that night; was it at these boys' house?

"You're Mr. David, the writer."

Panic closed me up. Everything was available and verifiable except for those hours between leaving my apartment and waking up at the train station. I woke up that next morning feeling so guilty and ashamed, and all of my hallucinations had been about my inadequacy and failure as a soul. Was there something I wasn't allowing myself to know?

As much as I liked to think I was an intrepid detective trying to put my life together and discover what went on during those missing hours, when a moped full of eyewitnesses stood right in front of me, I couldn't breathe. There had been children? I carried around a pair of white cotton panties, and I'd hung out with street kids?

"You have the wrong person," I said. "I've never met you."

The middle boy sucked his teeth and said something to the boy behind him, who stared at me in shock. The middle boy revved the bike, and the boy behind him spit on the ground as they took off, wobbling a bit before joining the constant meteor shower of traffic.

Soon thereafter, another Fulbright student in a different

part of town needed a roommate, and I left my tiny apartment under the elevator engine in Tarnaka. I came back only a few times to have dinner with Veda.

I never saw those boys again. I had the chance to find out how I spent the hours now permanently blanked from my brain, but I could not do it.

I was a coward.

Veda and I were at the dark restaurant full of mirrors. We ordered both rice and naan. I drank beer, and he drank water. It
had been over a month since I'd last seen him. I was nostalgic
for our old routine.

My new apartment was great; it was near a nightclub and a
bar that had "Heavy Metal Wednesday" nights. My roommate
was another Fulbright student. Eric was his name, and he was
getting his doctorate in history from Harvard. We talked basketball, and he had a TV, which aired the NBA playoffs. Eric
and I shared an office, and we'd chain-smoke and type during
the day, then at night drink beers and babble about Tim Duncan's perfect but boring style of basketball. Sometimes Eric's
girlfriend would visit, and the three of us would play Hearts
until four in the morning. Time just slipped out of my hands. I
was nearly done with my grant.

Over dinner, I thanked Veda again for being such a good
friend.

He pushed my thanks away with a crinkle of his nose. "For a
while when you were in the hospital, I did not believe what was
going on."

"You couldn't tell I was out of my mind?"

"You would seem perfectly lucid one moment and then be
very disturbed the next." Veda called the waiter over and ordered a Pepsi. Turning back to me, he said, "To be honest, I
thought you were faking it."

"You're kidding."

"I thought you were doing research on mental health hospitals." He held his hands up in the air, blocking out a headline.
"*Indian Psychopaths: An Inside Look.* I was very impressed with
you."

"When I pissed myself, you dropped that idea, right?"

"I thought you were very, let us say, *committed* to your plan," Veda said, delighted with his wordplay.

The steaming silver bowls of food came out. I piled rice and dal on my plate, then yanked off a piece of garlic naan.

"How would I go about finding Rajesh, the police officer from that day?"

"He works at the train station," Veda said in between mouthfuls. "You go down there, and if he is not working you ask when he will be scheduled next. That seems to be the easiest route."

"Should I bring him something?" I asked. "What would be appropriate for this kind of thing? Like, should I give him some money or flowers or something?"

Veda thought for a while. "The best thing you could do would be to write a note and give it to his superior, describing what he did for you. Then when he comes up for promotion, he'll have your note in his file. It would be very helpful for him."

I chewed and nodded. Veda was always very logical about these things.

The check came, and Veda tried to split the bill with me. I paid. We said good-bye in the parking lot. Both of us were on scooters. It was very dark, and Veda told me to be careful because there were crazies out on the street.

By the time I had finished my grant and left Hyderabad at the end of July 2003, I left a city that I had fallen in love with. Hyderabad was where I had spent the spring going to the nearby nightclub on hip-hop night, dancing with all the West African kids who were sent to India to study. It was the town where I had written a draft of a novel in my new apartment, and when the summer was at its worst I had become nocturnal and slept through the sticky heat during the day. It was where I had eaten lamb biryani nearly every night. The city where I had stolen a giant gold Styrofoam cricket ball from a bar, and when I had spent nights writing, I sat on the thing. It was where, on my birthday, I had accompanied Veda to a club to go dancing with the Russian exchange students and had gone to the bar twice to get Long Island Iced Teas for him, getting him drunk for the first time in his life. It was the town where, when things would get a little brittle at the edges of my mind, I would go down to the Charminar and listen to the arrhythmic sounds of the men who used hammers to beat silver thin.

I'd had other adventures as well before I left India. I had gone to Calcutta and attended a wedding, caught a stomach bug, and lost ten pounds in a couple of days. I had gone to Delhi and climbed a gate to escape a pack of dogs. I had gone to Mumbai and helped a friend who had broken her arm in a rickshaw accident. And I had eaten the dinner buffet at the lake-castle of Udaipur, which had been featured in the movie *Octopussy*.

I never went back to Goa.

During all of these travels, I carried the white cotton panties with me. I'd finally figured out that they were Anne's (the same brand, style, and size she'd worn when we were together in Ohio), a token of hers that I had taken with me when I originally left for my Fulbright. An e-mail exchange with Anne confirmed it. Now I carried them everywhere I went, stashed discreetly away in my luggage. If anyone had ever found them, I would've more readily admitted I kept them with me for masturbation instead of explaining the truth, which was much more embarrassing: I suspected they might've been part of what kept me safe during that night I couldn't remember. There was a slim possibility that they were magical, so I kept them close during my travels as a protective charm.

I had spent all of 2003 going out of my way to distract myself. I had ordered a suit made that was a copy of the one Bruce Lee wore at his sister's and mother's graves in *Enter the Dragon* (I brought my laptop into the tailor's and showed him the scene in slow-motion). I had posed as a journalist with a friend and attended the Gowd Brothers' Miracle Fish Cure, an annual event where each participant buys a pinky-sized live murrel fish and then stands in line. When you get to the front of the line, a small ball of yellow medicine is placed in the mouth of the fish, and then you swallow the whole concoction. It's supposed to cure asthma. The year I was there, with pollution in India so bad, it was the largest number of participants the Gowd Brothers had ever had.

I had stopped telling people about what had happened to me. I did my very level best to forget my amnesia, to ignore the anxiety, to banish the paranoia. I bought Valium at the local pharmacies for when it got bad and told myself that these pills were an extracurricular indulgence rather than a necessity. I had three different pharmacies that I circulated through just so no one would get suspicious.

I was fine.

The incident was the result of the Lariam, and the Lariam was out of my system. It had to be. Almost a year had passed. I was fine now.

I had to be.

The novel I'd written was funny and irreverent and had characters from India and central Ohio, and it revolved around a guy who went to India and got totally messed up, but it wasn't autobiographical. The story was told from his ex-wife's point of view. The man was near catatonic, and he babbled incessant nonsense, and in the book he is central to the plot but only felt peripherally in the scenes. If that main character was me—if I was lying when I claimed it wasn't autobiographical—then the thing I had written was about how I had figured out a way not to pay any attention to myself. In the end, the man gets locked away in a hospital and all of the other characters go on without him, their lives now free of his babbling.

When I was packing to return to the States, I pulled my desk drawer out and dumped the contents onto the bed. Among the detritus of pens, scratch paper, and matchboxes, I found a train ticket. It was dated October 17, 2002. It was for passage to the Vasco Da Gama Railway station in Goa. I had bought the ticket on October 9. So when I had left my apartment the day before with nothing but a pair of women's underwear, I had believed I was on my way to visit Geeta in Goa—carrying something from the woman I had left in New Mexico. When the Lariam pooled in my brain, nestling into my protein gap junctions, I was planning on cheating on my girlfriend. I was a divided self already.

According to the ticket, it was purchased on October 9, so I had spent a week and a half feeling conflicted before the chemicals in my brain went haywire. When I woke up, I wasn't the blank canvas I thought I was. The threads of shame and desire were already sewn through.

In the life I had woken up to, I found that I was often split between who I was and who I wanted to be. I grew up in small-town Ohio, but I wanted to be a world traveler; I went to small unheralded schools, but I wanted to compete with the country's best academics for a Fulbright scholarship; I was dating Anne, but I wanted to be the kind of guy who dated someone like Geeta. These aren't such unique fractures when compared to anyone else on the planet, but it was into these fractures that the Lariam nestled, and instead of being merely divided, it blew me apart from the inside.

2963

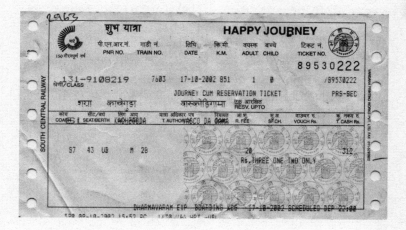

शुभ यात्रा **HAPPY JOURNEY**

SOUTH CENTRAL RAILWAY

150 गौरवपूर्ण वर्ष

पी.एल.आर.नं.	गाड़ी नं.	तिथि	कि.मी.	वयस्क	बच्चे	टिकट नं.
PNR NO.	TRAIN NO.	DATE	K.M.	ADULT	CHILD	TICKET NO.

89530222

131-9108219   7603   17-10-2002 851   1   0   /89530222

श्रेणी/CLASS

JOURNEY CUM RESERVATION TICKET   PRS-SEC

शशा   काक्तेमाड़ा   वास्कोडिगाम्पा   पुनः आरक्षित
RESV. UPTO

कोच	सीट/बर्थ	लिंग आयु	यात्रा अधिकार पत्र	रियायत	आ.शु.	हु.श्र.	वाउचर रु.	कु. नकर रु.
COACH	SEAT/BERTH KACHEGUDA	T. AUTHOR VASCO DA GAMA	R. FEE	SF.CH.	VOUCH Rs.	T. CASH Rs.

S7   43   UB   M 28   20   312

Rs.THREE ONE TWO ONLY

DHARMAVARAM EXP BOARDING KCG 17-10-2002 SCHEDULED DEP 22:00

539 08-10-2002 15:52 PC   1170 //63 HPT //IR

I never wrote the note for Josh. I never went to the railway station to shake his hand and thank him.

I left India and didn't thank the person who'd found me and been the first to help me. The one who started the relay of people handing me off, making the decisions in my best interest when I was incapacitated. I couldn't face him. I owed him too much. I shut the whole business out of my head.

Like I said, I was fine now.

PART FIVE

I know God will forgive me. No one could live with how I am feeling now. I know I will never forgive the bastards that gave me Larium. I am now the same as when I first had it—fully spinning can't even walk properly—the walls are moving. My head feels like someone let a box of ants in it, extreme pain in my head. I am fully losing it. What does the future hold—"psychiatric wards" no way. I know I've always been a little bit different even before I had Larium but since it first blew my brains apart and then settled down I have never been the same, always dazed and confused, always physically sick. I never thought this could happen to me. Sorry Mum, Dad

—Twenty-nine-year-old John O'Callaghan's suicide note,
two years after being treated with Lariam

The body has its own calendar. I became so adept at putting the previous October out of my head, I actually couldn't have told you the exact date I went missing. But as the anniversary approached, my body knew and was executing its own countdown.

At the end of July 2003, I returned to Ohio, spent a week there, and then drove to New Mexico with my dad and Sally the dog in my blue 1987 Toyota Camry station wagon. We stayed at a La Quinta in Las Cruces, apartment hunting during the day and drinking at the bars in the old adobe square of Mesilla at night.

I took my dad to the building that housed the English Department at New Mexico State University. I figured I could use him as a shield if anyone I didn't recognize came up to me to say hello. My dad would introduce himself to the new person immediately, and then at least I'd know a name. But my plan was worthless: the building was abandoned and the office locked. He and I walked the halls of the place, and we found a corkboard with departmental successes thumbtacked to it. On a square of newspaper was a picture of me and an article about my grant. It was written in May 2002. Dad looked up and down the hallway, then unpinned the article, folded it, and pushed it into my jacket pocket.

"They won't mind," he said.

We found a small house in my price range with a view of the Organ Mountains out the front door. It was just a half mile south of Mesilla, and I imagined a year of walking to the quaint and hard-drinking adobe town square every night for green chile enchiladas and cold cans of Tecate.

Dad and I returned to the hotel for our last night before we started moving things out of my storage shed and into my new house. It had been a sweaty day, so Dad jumped in the shower as soon as we got in the room. I pulled from my jacket pocket the article he'd stolen from the corkboard. I stared at the picture. I had a goofy smile that showed off my one screwed-up tooth. Me before Lariam. Could I send this guy and his dumb smile a warning? Because he had no idea what he was getting himself into.

I grabbed the hotel pen and blacked out the tooth, then blacked out another one. I drew glasses on the picture and added devil horns. Dad came out of the shower. He saw what I was doing and yelled at me.

"That was going to be for your mother."

"It's a picture of me," I said. "I get to do with it what I want."

"It was a nice picture of you, David. God forbid she should have one nice picture of you."

Back in the States, there were more things that fed my memory. During the week I spent in Ohio, I'd go on a run, and as I ran past a house, I'd remember the time I'd been at a sleepover in seventh grade there and had been locked inside a bathroom. I'd had a tantrum and torn the towel rod from the wall and kicked a hole in the drywall. Or when I was carrying groceries across the street back to my parents' house, I remembered having been a participant in a hog-calling contest when I was in elementary school.

Other memories surfaced. I remembered being in charge of my college's homecoming halftime show and having people dress up like Fidel Castro and Gandhi. They did a series of dirt-bike tricks before embarking on a choreographed dance to Prince's "When Doves Cry." I remembered the time I'd stood on a sidewalk in Chapel Hill, handing out absurd flyers I'd made at my boring office job about how the smell of Pepsi One made puppies go feral and eat children.

I'd bounce into things that I didn't remember from before that time and then fret over whether or not I'd have remembered it if the Lariam hadn't scrambled my brain. I was still afraid of the hollowness that I felt at the edges, afraid of what I didn't know that I had forgotten, afraid that it'd all happen again. But cigarettes helped with the anxiety. Alcohol helped with the insomnia.

I had one more year left of my graduate studies, and I put my books up on shelves found at a yard sale, bought a cheap laser printer, and got back to work. I worked on the novel that had the guy who went to India and came back to central Ohio insane and incomprehensible. I put in a lot of swearing and poop jokes.

The old adobe one-floor house I had rented had a big side yard with a pecan tree that I could tie Sally to. Inside there were four equal-sized rooms set up in a square, with a very small bathroom off the bedroom. When I told her that we were going on a walk, Sally did laps of the entire house. My landlord had grown up there, and when he came by, he would light my cigarettes for me. His dad had smoked when he was growing up. He seemed eager to have the smell of it back in the house.

The first week I was back in New Mexico there was a gradu-ate-student party at a stranger's house. One of the guests there was a white guy from New Zealand with tribal tattoos and dreadlocks. Someone assured me that I had never known him and that he was just visiting, so I didn't need to know him now.

This was something I was preoccupied with now — deter-mining who I knew and who I didn't. In the small three-year graduate program I was in at New Mexico State University, the class I came in with had graduated, and now I was in the class of the people who were formerly a year behind me. In the per-petual turnover of graduate programs, they knew the people the year behind them (who were all strangers to me), and then the brand-new class was full of certified strangers to everyone, even to each other. I spent most of the party trying to ascertain who was a genuine stranger to me and who just felt like one.

I met a couple I had never met before: Clark was a new stu-dent, a poet, and he and his wife, Tanya, were from Kansas. They were both good-looking and good-humored. Solid and stable people. I envied them immediately — their ease, their un-blemished lives.

Anne was there. She hugged me when I walked in, and then we orbited each other for the rest of the night, never seeming to end up in the same conversation cluster at the same time. She had gained weight since I'd seen her last. Her shoulders puckered over the spaghetti straps on her tank top. The weight looked good on her. She was from a small town outside of Green Bay and was genetically positioned to carry weight well.

I downed Tecates and powered through a pack of cigarettes. The hostess had a cello, and she uncased it and sawed some Bach out of it.

The night ended with the Kiwi noisily making out with one of the poets in a dark corner while I drank and smoked. I was a scarecrow who barely resembled a human, stuffing myself whole with alcohol and nicotine. Tanya and Clark both came up and shook my hand before leaving.

Sally was whining when I got home (how'd I get home? Did someone drive me, or was I dumb enough to drive myself? If someone else drove me, how was I going to get cigarettes the next day?). I let her out and stared up at the stars as she pissed. I leaned inside and grabbed a beer out of the fridge, downing it while Sally cracked pecan shells between her teeth.

I woke up on the floor of a convenience store to a trio of faces staring down at me. My back was wet. There was cold air pouring down on me from a freezer door that was held open with my head. Cans of Diet Dr Pepper lay scattered around me.

One of the faces, a black one with tiny moles on the cheeks, laughed as I opened my eyes. "You will never be a boxer, son," said the man. "I've never seen anyone go down that fast."

One of the other faces, a white female one with pimples coming up under a heavy layer of base, called me honey and asked me if I was okay enough now to stand up.

I got up slowly, my legs shaky. I was in a uniform: striped shirt tucked into brown polyester pants. I had a name tag on but couldn't read it.

"I don't think I've ever seen someone go down that hard," the black man continued, the glee in his voice too evident. "That door hit you, and you hit the floor."

The woman with the makeup pulled me aside. "I don't think you have a concussion or anything. Go wait in the office and collect yourself." She turned her wrist over to check a thin silver watch. "I can't send you home. Just take your lunch break now, grab a fresh shirt from the pile, and take it easy. Okay, sweetie?"

I gestured to the mess behind me: cans of Diet Dr Pepper were smashed open and fizzing out onto the linoleum, a whole display shelf of Pringles was crashed down on the ground, the red tubes scattered down the aisle. "Did I do that?"

"You can take care of it when you come back. I'll get Reggie to finish the stocking." She leaned toward me and kissed me on the cheek. "Remember, I'm going to need you to pick up the kids from soccer after you're done here. I've got court."

Puzzled, I nodded and went into the room the woman gestured to. It was a tiny closet, just big enough for a desk and a chair. The desk was stacked with four monitors, which showed in scratchy black and white the goings-on of the store, the gas pumps, and the living room of my parents' house in Ohio. I couldn't find the light switch. I sat in the dark in front of the monitors' flickering light. I could feel an emptiness in my head, in my chest, in my feet. I was hollow.

I woke up from the dream, shuddering in a pile of puke. I'd blacked out, passed out, and vomited. I was hungover, and my chest was so tight that it wasn't letting me take a breath in. I was sweating, shaking, and nauseated. Sally came up and licked my hand. She leaned into me, a generous affectionate quirk of her breed. I grabbed onto her fur and hugged her and wept.

Over the next two weeks, I had several more dreams in which I'd wake up as other people. I was never anyone extraordinary. I awoke as the guy at the convenience store most of the time, although once I was a tax attorney. All the pyrotechnics of the previous year's hallucinations had drained my brain of its imaginative sizzle. My dreams were boring, each one having a distinct moment of me sitting in a chair in the dark feeling empty. I could wake up as anyone. I was interchangeable as a personality; what I thought was me seemed to be inconsequential and could be used to fill up any life. There was nothing special about me; I was a fungible soul.

Who I was never mattered in these dreams. At the end of them I'd always end up by myself in a tiny room: storage closets, confession booths, walk-in freezers. Alone and stuck somewhere dark and quiet. I'd wake up shaking and crying after each one. I called them banal-mares. I didn't tell anyone about them. I didn't want anyone to think I was going crazy again.

Through an odd confluence of events, I ended up using the office belonging to one of my professors. He was on sabbatical and
I had a key, so instead of bunking with four other graduate students in the tiny broom closets they called offices, I hung out in
his office, with its shelves of signed first editions and his fancy
computer. I retreated to that office often. It had a little partition
that separated the front part of the office from the back part, so
during my mandatory office hours, I could have the door open
and hide back in the computer area. The lazier students would
look in, not see me, and leave. It was the perfect way of not
doing my job while doing my job.

One morning I was hiding behind the partition. I had a hang-
over and had woken up to a panic attack. There was something
going on with my right ear. An infection was building up in
there and gave me stabbing pains, one of those pains where you
can feel the exact diameter of the infected canal. I didn't have
health insurance, but I could go to the student clinic for free. I
just didn't want them prescribing me medicines I couldn't af-
ford.

The phone rang, and before I thought better of it, I answered,
cradling it against my uninfected left ear.

"Is this Dave? Dave MacLean?"

I didn't recognize the voice.

"I heard you were back. It's Jacob."

I didn't recognize the name. His voice was charged with
excitement at being able to speak with me. We must've been
friends. I asked him how he'd been.

"So Bonnie told me that I had to call and tell you this. You
remember that paper I wrote? The one on cancer?"

He was a former student. For a second I wondered: had my
brain not been shaken like an Etch A Sketch by Lariam, would
I have remembered a former student's paper? They turned in so

many of them. I told him that yes, of course I remembered his paper on cancer.

"Well, I got it," he said. His voice broke a little. There was a manic edge to his excitement.

"The paper?" I asked.

"No. I got cancer," he said. "I lost a testicle."

"Oh."

"I thought you'd think that was funny." Jacob's voice was crackling now. "Bonnie and I laughed a lot about it."

Jacob ended the call by asking me to come down to the Double Eagle, where he waited tables. He promised he'd get me half-priced appetizers and we could catch up.

It was nice to know that I was the kind of teacher that students liked and in whom they wanted to confide. It was one of the talents I found that I had: I was a good teacher. But I was also the kind of person who my students were sure would find their cancer hilarious.

One night, I met Anne out for drinks. We had many. Then I went back to her house to check out her computer, which she said was acting all weird.

I woke up at two in Anne's bed. She was wearing white cotton underwear. I told her I was scared. I told her that we couldn't do this anymore.

She reminded me that we hadn't done anything past hold each other.

Crying, I told her I was messed up and she was dodging a bullet.

I drove home. The pain in my ear felt like there were bits of broken glass in there. At my house, I stood outside smoking and listening to Sally cracking open pecans. The giant lumbering shapes of the Organ Mountains were visible even at night. You looked out, and where the stars weren't, there were mountains. Sally pushed my hand with her nose, and I took her inside and dumped kibble in her bowl. Then I drank myself to sleep.

The second week of October, the pain got so bad that I finally went to the doctor and begged for a path to wellness paved with cheap generics. She wrote me a prescription for antibiotics for the infection and OxyContin for the pain. She told me I shouldn't drink with either of the medications. OxyContin was a narcotic, she'd advised me, and when mixed with alcohol it could be deadly.

On October 16, 2003, the clock ticking inside my body went off. I took three of the Oxys, drank a bottle of wine, and had a tsunami-sized panic attack. My body knew the one-year anniversary was here. My chest was so tight that I could barely breathe. My heart thrummed at a hummingbird's pace. My thoughts whirled around. I was drunk. I was numb with painkillers. I paced around the house. I shoved my fingers deep into my throat and threw up everything until my stomach convulsed air.

I called my parents. I opened another wine bottle. I called my sisters. I took another pill. I called Jon and Melissa, my friends from college. I had a cordless phone and spent hours spinning through the rooms, spending money I didn't have on long-distance calls. I needed to have reasons not to kill myself.

I asked my family and friends, why should I try and do anything if I could just lose it all again? It was futile. I'd ramble on about Sisyphus, the mythic figure whose punishment in hell was to push a rock up a hill, and then right as he was about to get it to the top, he'd slip, and the rock would roll down to the bottom again. If Sisyphus had the choice, he would stop pushing the rock, right? Who in their right minds would continue to push the rock up the hill voluntarily when they knew that they were going to fail? I had spent the last year pasting together a passing resemblance to myself, and at any moment I could wake up and not know who I was and have to start this all over again, without even the memory of how I had done it. Why not commit suicide, if this was the case? I'd get talked down by each phone call, ending it by assuring the person that I was all right and hanging up the phone.

Then I'd punch in someone else's digits. I was out of wine and had switched to beer. If I could lose everything at any moment, why not just die at a moment of my choosing? I barked into the phone. What if it wasn't the Lariam? I asked. What if I was wired wrong? I didn't want to have to read all those books again. Hurt somebody like Anne again. Go through this carousel of shit again. The people on the other end of the phone told me it might look dark right now, but it was going to get brighter soon. I replied that it'd get dark again after that brightness. People on the other end of the phone said that there were those in the world who had it much worse than I did, and that I, of all people, should know this. I felt so out of control that suicide was attractive. I was told that what I lacked was a global historical perspective of suffering. I wanted a gun big enough to guarantee an eradication of each and every one of my brain cells.

I was drunk, inconsolable, and my fingers stunk of vomit. Life felt like a too-long race, all spent running in wet concrete, each year a little deeper in: toes, knees, pelvis, chest, neck, death. I'd seen God, hadn't I? I'd seen the earth in four dimensions, been to a meeting of angels. I had been shown that there was just darkness and space and failure. If my hallucinations meant anything, they said that I wasn't getting into any afterlife. Suicide seemed the logical choice.

Sally walked with me all through that night, her wet nose pushing at my hand if I stopped pulling at her ears or scratching her head. *Pet the dog*, a voice inside of me yelled. *Pet the damn dog*.

I had a billion voices in my head, all clamoring for attention. Everyone I loved was in other time zones, and they were getting tired; they had work the next day. They told me that things would get better. My brain was indefatigable. The noise of it all was deafening. The chaotic crest of an orchestra tuning up combined with the sonic dissonance of an avalanche of radios, all caught between stations. I drank some more.

I noticed suddenly that the pain in my ear was gone.

I missed it.

The pain had been something concrete, something that tied me to this world. I missed the pain that I'd started the day complaining about.

"Isn't that just like you," I said.

I woke up on the foldout bed of a sofa in a living room. Not mine. Through the sliding patio doors I watched two small brown birds ducking their heads into a puddle, raising their beaks, shivering themselves dry. Outside, everything was so green. A machine rattled and exhaled in another room, and all at once the world smelled of coffee. It was October 17, 2003, and although I had no idea where I was or how I'd ended up here, I still knew who I was.

And I had a massive headache.

I was still in my jeans and shirt, socks, and shoes. There was a chunk of ice in my chest where the anxiety had been. I stood up, made the bed, and then folded it back into itself. It collapsed neatly into the sofa with a thunk.

My friend Jenn came in through the front door brushing dirt off her shorts; she had a paperback creased between her fingers and a giant coffee mug. I was at her house. It was a half mile away from mine. I'd been at this house before. I knew where I was.

She refilled her mug and poured one for me, and then we went and sat outside. Pine trees surrounded her house, and it was so isolated; it felt like a place burrowed deeper into the world. Across the street there was a pasture, and two horses calmly grazed. I wanted to feed them something, to have them eat from my outstretched palm and feel their floppy lips and the hot breath from their nostrils. Instead I sipped the bitter coffee and sat there for a half hour by my friend and watched the horses nicker and graze. I stuck two cigarettes into my mouth, lit them, and passed one over to my friend.

We said nothing. The ice in my chest was heavy, but the pine trees and the coffee and the cement slab under me all presented themselves as miracles. My car wasn't there. Jenn had come and picked me up at my place in the middle of the night. She was a miracle as well. The fact that my brain hadn't reset again, that I wasn't hallucinating or strapped down to a table, was disconcerting to me. The fear of going crazy isn't the same as going crazy, though it feels similar. Being crazy is easier than being afraid of going crazy; in fact, it's enviable. The hospital was awful, but there they put straps on me to keep me from injuring myself. If I was crazy, everyone else would make decisions for me, for what was in my best interest. I had no such protections anymore.

Except for Jenn and all the other people who listened to me that night as my brain unspooled. The kindness of a place to sleep, a cup of coffee in the morning, and silence as we watched the horses graze: unfathomable gifts.

The smell of the trees on that day is still with me.

I started going to a shrink. She was the cheapest shrink in Las
Cruces. I was spending a lot of money on cigarettes and alcohol, and didn't have much cash left over to spend on the reasons I was smoking and drinking so much. Here were two substances that allowed me to control the way I felt. Nothing else in my life gave me the kind of comfort that that control gave to me. I was putting packs of cigarettes and fifths of bourbon on my Discover card and then paying only the monthly minimum — treating mounting debt like another utility, something that was an unavoidable part of keeping myself sane.

The shrink I went to was in her early sixties and was partial to lengthy sheer scarves, which floated around her when she walked. She worked with people with post-traumatic stress disorder, which she diagnosed me with during our first phone call.

She apologized to me on the day of our first appointment because her dog had been needy that morning, and she hoped it was okay that he was going to be with us.

I told her it was fine, that if I could, I'd take Sally to restaurants and to the movies with me.

The shrink laughed and opened up a side door from which the dog popped out.

Her dog's name was Allegra.

"Like the medication?" I asked.

"It's Italian for happiness." She grimaced.

The dog was black and white, about forty pounds, and was missing its left front leg. Allegra hopped over to me and smelled my pants as I rubbed his head.

As the shrink settled into the opposite couch, she explained that she had an intake form with a list of questions to go over, but first she wanted me to explain in my own words why I was seeking help.

I told her the story of waking up in a train station. By the second sentence I was crying. Fifteen minutes in, and I was sobbing convulsively. By the time I finished, the hour was over. We saved the form for the next time.

Las Cruces is famous for its green chiles, or, more specifically, it's adjacent to a place called Hatch that's famous for its green chiles. Long, spicy, and sweet things about half the length of a banana. When the harvest came in, everything ground to a halt. The peppers are roasted in black propane-fired tumblers, usually a little bigger than the size of the man who stands next to them spinning the black handle attached to the drum. Grocery stores, gas stations, farmers' markets, restaurants, and bars would each put a roaster out in their respective parking lots and sell the chiles by the bagful. The air in Las Cruces in the fall is thick with the sweet and burnt smell of them.

The best restaurant for green chiles in town is out of town. You drive down an old road that used to be the main route between Las Cruces and El Paso. It takes a while, but right when you think you're lost, you end up at Chopes.

Chopes is two buildings: one is the bar, and one is the restaurant. Both are tiny and loud. The front parking lot is stacked with motorcycles, and the insides of the buildings are filled with black leather–clad giants. The bar serves quart bottles of beer that look proportional in the fists of the giants. A typical wait for a table on a Friday night is well over an hour, so you find yourself perched on one of the impossibly high stools at the bar, and you can end up so drunk that you can't imagine why you ever wanted to eat in the first place. But the phone is always ringing in the bar, and it's always the restaurant with another open table; your name gets called, and you head over across the gravel parking lot. Outside, it's colder than you expected, and the first chilly draft of air lets you know how drunk you really are. When you start to talk, you realize that you're still taking into account the bar noise, and you find yourself yelling at your friend until you catch yourself, and you both bust up over how the night is spinning predictably out of control. You zip your sweatshirt a little snugger when you feel yourself being sized up by the giants smoking astride their machines. There's a giddy slickness that comes over you when you realize that you might end up in a bar brawl with bikers. You want to push one of those metal monsters over just to hear its headlights pop in the gravel, just to have one of those giants punch you right in the face.

The restaurant's screen door bangs behind you, and there's a line of people in a thin yellow hallway. Your enthusiasm for the meal is coming back and teaming up with your drunkenness; it makes you loud again, but this stops quick with a look from the tiny woman at the front. She's old and stern, and she'll only seat you if you shape up. You leave the hallway and realize that the restaurant is somebody's house. The kitchen is in the back and looks like your kitchen, cramped and prone to smokiness. There are tables in what used to be the living room, tables in what used to be the dining room, and a long row of tables shoved together for your party's benefit in what used to be a den, the whole place done in wood laminate paneling. The restaurant is staffed exclusively with tiny, disapproving women.

What you order matters very little. Every plate that emerges from that little kitchen is a uniform beige. Plates of beige, the entire beige rainbow. You order another quart of Tecate and dive in to the chips. Someone at your table is bound to put his hand to his mouth and claim that the salsa is too spicy for him, maybe even indelicately spit out the chip into his paper napkin. The beer comes; the food comes; people eat half their plates before they realize their orders are switched. Ha ha ha.

You argue about the check. Everyone is short. Everyone is cheating and forgetting to add a tip. Somebody is sulking about how they shouldn't have to pay since their food was half-eaten by someone else. The scowling women shush you.

You're outside in the night air. You want to throw a rock at a parked motorcycle. You want to run down the center of the empty street. You want to wrestle. Why doesn't anyone else want to wrestle?

Now you're driving, the road slaloming between giant lurking dark figures of pecan trees. Their outstretched arms press into the side of the road, reaching for you. You're drunk, but you have the window down, and you asked your buddy to drive in front of you so you can steady yourself with his brake lights. You watch them like subtitles. The cold air stabs clarity into your eyes, your blinking eyes; you keep blinking your eyes.

The dog is whining when you get home. You're an awful dog owner. You apologize to her as she pisses in the front yard. You tell her you know that she'd be happier somewhere else, that you don't deserve her; you shove your face into her fur and apologize. You smoke a cigarette and decide to walk to the bar in Mesilla. It's not that far, and you can get there before last call. The cars on the highway drive too damn fast, and they honk at you walking on the shoulder, the Doppler effect of their horns and tires ricocheting into the valley. You pass the gas station and decide to buy cigarettes there instead of at the bar because it'll be cheaper. In the gas station the fluorescent lights belong to another planet. You buy two packs of cigarettes, and the person behind the counter is amused with you, the way you fumble with your wallet, the way you press the pen into the receipt too hard and tear it as you sign more debt to your name. At the bar you don't see anyone you know and the music is awful and there are all these women in burgundy bridesmaid dresses and you order a double rum and Coke for some reason and it's last call and you can't get the bartender's attention and you leave without paying because they were playing awful Jimmy Buffet Parrothead music and that's what they deserve and the clerk's look from the gas station has stuck with you and the dog's whining has stuck with you and every passing honking car horn has stuck with you. You try and talk yourself into some tears. Crying will knock all of this out of you; you feel like an emotional catharsis on the shoulder of the dark highway will make the world feel real again, will stop everything from feeling as distant, plastic, and prepackaged. Break this awful feeling like a fever. But no tears come. That hollow feeling is there instead, the one that makes you want to tackle light posts, jump through plate glass windows, lean into the road and get clipped by a Jeep Wagoneer, cause some disruption in the world, end up on a gurney, get strapped down in the back of an ambulance, let someone else monitor your fluids and check your vitals, let there be phone calls on your behalf and meetings with a quorum of the concerned who claim that they were unaware that it had gotten this far out of hand, that you had gotten this far out of hand, that you obviously need the clear road as defined by the handling of professional caring others. Because all of these decisions, these decisions of cigarette brands and combination plates and quart beers and driving drunk, it's too much for you. The warm womb of a hospital room, the sharp scrape of a plastic spoon hitting the bottom of a plastic bowl as someone feeds you curd rice, let you be born again again, hit the button and be reset again again. You got it wrong this time.

You get home safely. You're not lucky enough to get into an accident. There is nothing in the fridge and the bookshelf is full of things you haven't read and you're too exhausted to not turn on the TV and the dog licks your hand apologetically. You're not going crazy. You watch for it all the time. But you're not. You're drunk and not going crazy, and continuing on in the world of the sane is harder than you thought.

You don't want to want to die.

Alan Moore wrote a story about Superman called "For the Man Who Has Everything." It's Superman's birthday, and Batman and Wonder Woman have come to his Fortress of Solitude in the Arctic to give him their gifts. They find him standing catatonic in his trophy room, a box and wrapping paper at his feet and a strange bit of plant life lodged in his chest, its vines encasing his bulletproof torso.

Superman is an immigrant, one of the last surviving members of a doomed planet. He's the rare superhero who has two secret identities, one nestled inside the other. He's Clark Kent, Kansas farm boy turned Metropolis reporter, and he's also Kal-El, born on the now dead planet Krypton. In addition, he is also Superman. He is three different people, constantly moving between costumes.

Birthdays for Superman are bittersweet occasions. If his home planet hadn't been destroyed, he'd be an ordinary, non-superpowered citizen of Krypton, just Kal-El. In order for him to be Superman, he had to be sent some million miles away from his family.

Superman has friends, but most of them wear masks. Clark Kent has friends, but he has to hide his real abilities from them. He has three identities, and none of them seem to be who he truly is. He's the most famous man in the world, yet he chooses to spend his birthday in his Fortress of Solitude, an ice and steel structure he built for himself in the frozen wasteland of the Arctic.

Now one of the gifts that he has opened was a trap, and a parasitic organism has attached itself to him. Batman and Wonder Woman can't remove it without killing Superman.

Suddenly, the enemy reveals himself. He's the standard thuggish oversized alien would-be world conqueror that Superman has dedicated his life to fighting. His name is Mongol. He knows he could never beat Superman in a fistfight, so he developed a trap using an alien plant. What the plant does is feed on the dreams of its victims. It induces a coma in its host, giving him visions of his greatest desire in order to keep him placated. While Superman is out of the way, catatonic with hallucinations, Mongol can finally conquer Earth.

In Superman's hallucination, he's Kal-El, living on a Krypton that never exploded. Kal-El comes home one night after working late at the lab. The lights are off in his high-rise apartment, which overlooks the gorgeous Kryptonian skyline. Kal-El calls out greetings first, and then apologies to his wife, to his children, for being so late. All of them go unanswered. Then, just as the edge of malice starts to set in, the lights all pop on at once, and—surprise! Everyone is gathered in the living room, cheering Kal-El a happy birthday. Kal-El makes the rounds of the room, hugging coworkers and family members, pausing to kiss his wife. The party ends, and everyone leaves. Right before they make love, Kal-El's wife makes a comment about how he's been so distracted recently.

As Superman hallucinates, Mongol starts to fight Batman and Wonder Woman. Batman is no match for the fisticuffs, so he ducks out and tries to figure out how to detach the plant from Superman. Wonder Woman gives her best. She and the villain trade incredible blows, which send each of them careening through the metal walls of Superman's fortress.

In Superman's mind, Kal-El takes his son through the streets of Krypton. His son is curious and quick, peppering him with questions. Kal-El is still troubled. None of this seems right to him. His wife. His son. They are all he could have ever wished for, which makes them suspect.

That day he has taken his son to the crater left behind when the city of Kandor mysteriously vanished. Kal-El brought his son here to teach him a lesson about loss. Here was a city gone, all of its secrets forever kept, which meant that they weren't secrets anymore, just absences. A crater is an absence you can see.

A fog rolls in. Kal-El turns to his son and tells him that he loves him, but that he knows that he's not real, that none of this is real. The son starts crying. He begs his father not to say such things.

Superman tells his son that he's a hallucination.

The plant releases his grip on Superman's chest, and Batman places the plant back in the box. Released from his stupor, Superman beats Mongol, then places the plant on Mongol's chest, imprisoning him forever in his own personal fantasy.

Superman had to redestroy his home world in order to come back to reality. He had to turn his back on the loving wife he never had. He had to leave his son crying at the edge of a crater in order to return to himself. He had everything he wanted in his hallucination, but in the name of reality, he had to reexperience all the pain of his orphaned, marooned life on an alien planet. Even a comic book character has the sense to recognize the validity of the real over the imagined.

I tell you this because there was something lovely and horrid in my hallucinations. The emotions stirred within them were real, even if the images weren't. I was denied access first to my loved ones, then to the upper ranks of the angels, and finally to transcendent knowledge — ultimately rejected by God himself. I saw the breathtaking depth of a world in four dimensions. These things never existed outside of my skull, and I felt hollowed out by them. I knew better, and it didn't matter.

My hallucinations left me feeling like the inside of my soul had been flapped out for the world to see; the shame I'd carried through my life had bubbled out and been exposed to the air, and now it wouldn't recede. I felt like my veins were on the outside of my body.

Halloween came, and I covered my face and arms with chalk dust and blue body paint, wrapped the rest of me with aluminum foil and saran wrap, and duct-taped a digital clock to my chest: I was a cryogenically frozen man. The party was in the backyard of one of the guys in the program. He was dressed as John McEnroe, in tiny white shorts, a tight white polo, an Afro wig, a headband, and a prosthetic cock that extended out of his shorts by ten inches. He greeted every guest by grabbing his racket and dropping down into the ready stance, the cock swinging wildly between his legs.

There was a bishop, a naughty librarian brandishing a whip, a Marilyn Monroe, a woman in a pink wig who'd made a bustier out of duct tape, a woman in a disco-era catsuit, and a very pregnant woman who'd made herself up to be Tony Soprano. Anne came in a dress onto which she'd stitched a dozen plastic lizards.

Clark, the poet from Kansas who was half of the incredibly good-looking couple I'd met at the beginning of the year, came to the party by himself. He said his beautiful wife, Tanya, wasn't feeling well. She had been absent from several parties recently. When I got there at nine, he was shotgunning cans of Coors Light and weaving around with a bottle of Jim Beam.

There was dancing. Magnetic Fields, Jay-Z, Neutral Milk Hotel, OutKast, Kanye West, the Shins. As I danced, my costume peeled off me. I had left my car at someone else's house, so I borrowed a shirt from well-hung McEnroe and sat down for a while to cool off. A woman came up and sat next to me. She had showed up with her husband and was wearing a pair of cat ears she'd probably picked up at a gas station. I'd met her husband earlier in the year. He was short and had incredibly thick skin, like a pumpkin, that advertised every nick and scratch he'd ever received. I never heard him voice an opinion everyone didn't already agree with.

His wife and I started talking. She was drunk and immediately started pulling cigarettes out of my pack without asking and divulging intimate secrets. Within five minutes, she'd told me that she'd cheated on her husband while they were engaged and that it was a testament to how great a guy he was that he married her anyway. I remembered something that Geeta had said and told her that monogamy was the appendix of modern culture, primed to burst. I was a little drunk, too.

"You don't respect women," she said.

"What?"

"It's something I noticed about you when we first met two years ago. We were at a dance, and you were dancing with Jenn, and you dropped her." She took a drag off the cigarette she had bummed from me and continued, "Instead of helping her up, you turned your back and kept dancing."

"I was probably drunk," I said. I had no memory of the event, so I couldn't refute it. Jenn was the woman whose house I awoke at after my anniversary night of panic attacks and drunkenness. She didn't seem to hold a grudge.

Cat Ears blew smoke out in a thick rope and shook her head. "And then when this stuff with Anne happened, it wasn't a surprise to me since I knew that you had issues with women."

"I don't respect women?"

"Take my husband. He and his brothers treat their mother like a queen. She doesn't have to lift a finger when they're around."

"You have no idea what happened between me and Anne," I snapped at her. I was swimming in the borrowed shirt, and the saran wrap on my legs was starting to itch. How the hell was I going to get home?

"I saw you drop her and keep dancing. That told me everything I needed to know." And with that she pulled out three cigarettes, tucked them into her shirt pocket, and threw the empty pack at me.

One morning as I sat outside watching Sally chase after the tufts of cotton that blew off the field behind the house, a truck full of men with survey equipment showed up in the empty lot across from me. They marked lines in the dirt. They drove rebar into the ground at specific points and then strung twine between these points. Just like that I was separated from the mountains by the outline of a fence.

The birds found the fence quickly and snatched the leftovers of the workers' lunches, perching on the taut twine while the men set about digging trenches and erecting wooden forms.

The next day a cement mixer showed up, and the men used hoes and rakes to coax the gray slop into the forms, then smoothed the top of the forms with flat pieces of wood. By the end of the week the wooden forms had been broken away, and the new walls of the fence were painted pink to look like adobe. It was five feet tall and obscured my view of the foothills.

At night, I'd grab a Tecate from the fridge and walk Sally through the newly asserted compound. Within a week, a cul-de-sac had been marked, then graded, then paved. Driveways to nothing started showing up. Then one night when I came home and put the leash on Sally for our walk, we found giant yellow machines sitting curled up, hulking, and quiet next to giant holes, as if they had nothing to do with them. Each drive-way had an adjacent hole. The new fence now looked like it had been built so that these new holes could be protected. I threw my empty can down into one of them. Sally peed on one of the machines.

The world was moving so quickly. I felt like I'd never be able to catch up.

Anne stopped talking to me. According to mutual friends, she was now on a new regimen, swimming laps every single day. To see her on campus would be to see the damp outline of her suit wetting through her outfits. She also carried around an enormous mug with a crinkled straw poking out that she was forever sipping from. Friends said she wasn't eating, but she was drinking over two gallons of water a day.

I received two packages in the early days of November. One was a knit scarf from Ariel, my ex from years back. The note with it said that she hoped I was doing well and that I could call her if I wanted. The other was a beat-to-hell copy of *The Wonderful Story of Henry Sugar* by Roald Dahl and a long handwritten letter from a woman acknowledging her part in the problematic relationship that we had had. She wanted me to know that she valued the time we had before things went bad. The book, she said, was hers from her childhood. It had always been her favorite, I reminded her of parts of it, and she thought I should have it.

I had no idea how to respond to the scarf or the book. I remembered the women. I had pictures galore of Ariel, and there were a few snapshots lying around of the other woman.

Reconstructing our relationship through old e-mails and by way of stories from friends and family, I found out that while I was dating Ariel, her cousin, with whom she had been exceptionally close, died in a car accident. We were dating long-distance at the time. She was in Boston, and I was in Las Cruces. She cheated on me. I cheated on her with the woman who sent me the book. ("Right after her cousin died?" I asked. "Yeah," my friends and family said.) Ariel and I spent the next summer house-sitting in Chapel Hill, and it sounded awful. We'd have friends over for dinner parties and end up screaming at each other in front of everyone. She went back to Boston and I went back to Las Cruces, where I applied for the Fulbright grant. I called her that November and broke up with her.

And that's when I had started dating Anne.

Back at the shrink, I had some new thoughts and wanted to bounce them off her. I felt that memory was a cultural construct, all of it preshaped by commerce. In my teaching, my students were much better able to talk about movies and shows than they ever were about themselves and their experiences. My experience with amnesia was most real to others when I talked about pop culture antecedents. I wheeled my hands in the air, which is something I found I did when I was trying to say smart things. The only way I could get people to understand the experience of waking up to being me was to talk about the movie representations that weren't like my experience. In order to talk about me, I had to refer exclusively to the not-me.

The shrink listened to me. She folded her hands. Kept them in her lap.

I continued. I talked about oblivion and how we only know it through the doorways, the bottomless pits that heroes balance above, the sarlacc pit in *Return of the Jedi*, a volcano holding a hole into which sacrifices are thrown. My hands weaved intricate designs in front of me as I spoke. I spun my wrists. I pointed to pockets of air as if my abstract concepts existed in that very space. I wanted the shrink to understand that nothing is always defined by something; that in order for me to understand my experience of nothing, I've had to become gnarled and ugly at my borders.

I was really on a roll.

My heart was beating fast, and I was flush with the sheen of ideas. This was real progress. I started to think that the shrink was going to go home and write papers about me. I wondered if I'd get any credit. What'd I say — "a volcano holding a hole" — a hole that is held? I'd have to remember that for later.

I went on for nearly the whole hour. I was giddy with the brilliance I tapped into. I was amazed that I was paying for this.

She should have been paying me. I quieted down and sat on my hands. I was prepared for a summative statement from her; it would come out like applause.

She folded her notebook and shook her head. "You've done a good job of saying everything but how you feel," she said. "Sadness isn't something you get to get out of by being smart. You don't get to outwit this. You will have to deal with the pain at some point."

I went home for Thanksgiving. Every time the plane pitched up and down, my chest tightened up. I wanted a drink. The movie they were playing was *The Core*. I decided not to watch it. Instead, I concentrated on making sure I continued to breathe.

Leaking out of the headphones of every movie-watching passenger were the sounds of explosions and metal-on-metal grinding. The plane bumped up and down. I pressed my call button, but the turbulence was too extreme, and the flight attendants were all strapped in as well. We kept pace with the storm, and the flight would suddenly drop, and my stomach would enter my throat. I flipped open the window shade. It was all darkness out there. Chunks of cloud flew past the window, the little light on the tip of the wing bipped on and off, and I could see the wing bending as we flew. I kept waiting for it to snap off. Out of the corner of my eye, I watched Hilary Swank and Aaron Eckhart pilot a digging machine to the center of the earth in order to deliver a nuclear payload to restart the suddenly dormant core. The cabin of their digging machine bucked and spun as they navigated rivers of magma and fields of diamonds. I watched the wing bounce up and down. The plane took a big bump. All of us lifted out of our seats, our seatbelts straining to keep us all in.

My call button stayed lit, but the flight attendants stayed in their seats the entire trip. It was three hours of rocking and pitching and wailing and a sweaty Hilary Swank gritting her teeth, pushing the digging machine deeper and deeper into the earth. The sweat ran down my back, and I pushed my palms down my pants again and again to keep them dry.

The clouds tore by like clawed cotton as we descended. We seemed to stay in the clouds for forever. When the city appeared, we were right on top of it. The dim Columbus skyline was surprised to see us emerge.

I woke up not knowing where I was. The room was tight. The
shadows were strange. Rows and rows of floating passengers
were all around me, two tiers of them lined up and facing the
same way. I was damp and on a mattress in the middle of these
rows of people. None of them looked at me. The moonlight
hit their outfits but didn't reveal their faces. Something hairy
brushed my hand, and I jumped.

I hit at the walls, and a light came on. My parents' dog looked
at me worriedly, giant saucer eyes wondering what had disturbed our sleep.

I was in Ohio. My parents had turned my old room into a
closet, and I was sleeping staring up at double rows of clothes
on hangers. This kind of thing happened to people all the time,
I told myself. This wasn't prodromal of anything except ordinary life.

In the corner of the room were boxes full of comic books.
My comic books. Thousands of them. I had left them here with
my parents when I went away to college, and they had stayed in
that corner ever since. I flipped through a few of them, hoping
they'd get me drowsy enough to slip back asleep.

It was five a.m. when it became clear that I wasn't going back
to sleep no matter how many old comics I read, so I went downstairs. My mom was up, watching one of the network morning
shows while using a spreadsheet on her laptop to plan the day's
oven schedule for Thanksgiving dinner. I fixed myself some
coffee and sat down with her.

During a commercial break I asked my mom if I respected
women.

"You were raised in a house full of them. That's for sure."

I told her about the woman at the Halloween party with the
cat ears and what she'd said about my dropping Jenn on the
dance floor and how that was illustrative of how I treated all

women, including Anne, including the way I treated my own mom.

She closed her laptop. "David, the real problem with her statement is that it makes it seem like you respect men."

"So I'm really just a jackass then?" I asked.

"My son, the equal-opportunity jackass."

The new semester came, and Anne and I ended up working the same shift at the campus writing center. It was three hours, twice a week. Industrious kids with bad grammar made appointments with either one of us, and we'd walk them through subject/verb agreements and the differences between *your* and *you're*. There was a lot of dead time, and Anne and I would spend it at our respective terminals, silently surfing the Internet. She was skinnier than ever. It was early February in Las Cruces, but she wore flimsy sundresses that showed the stark anatomy of her clavicle. From friends I had heard she was still swimming every day before she came to school in the morning, and three days a week she swam again before she went home. She carried that giant mug of water and trailed chlorine smell through the halls.

Sometimes, Clark would come and sit with us on slow days. It was early in the semester, so not many students were scheduled. Clark's wife, Tanya, had left him for another woman. I envied Clark; he was up-front about how miserable he was. He was drunk nearly all the time. At bars, Clark would bitch about his lesbian ex-wife, and some smart-ass would say that the exact same thing had happened to Ross on *Friends*. The worst heartache of his life, and a TV show had gotten there first.

Clark's ex-wife worked at the used bookstore in downtown Las Cruces. The place was massive, an adobe labyrinth with meandering shelves that branched off into other rooms. It smelled of moldering paperbacks and microwaved enchiladas and was easy to get lost in. Clark's ex-wife was dating a woman who also worked at the bookstore and who wore black T-shirts and carried a wallet with a chain on it. It wasn't out of the ordinary to stumble on the two of them making out in Contemporary American Poetry M–Z.

Clark, Anne, and I would sit there in the writing center

watching the sun etch down the sky before hiding behind the mountains. The drunk, the anorexic, and the amnesiac. All of us in our own little separate pods of sadness.

I don't know when the two of them fell in love, but I was probably there to see it happen.

The insomnia never stopped. I'd go a week and have perhaps ten hours of sleep tucked in here and there in half-hour segments. I didn't want to go to a doctor, didn't want to tell anyone anything anymore. I didn't want more pills. I didn't want to be diagnosed as crazy. Again.

A video store up the street from me went out of business and was selling off its stock for two bucks a pop. I took a cash advance on a credit card and bought fifty or so tapes.

I watched *L.A. Confidential, The Terminator, Miller's Crossing, Unforgiven,* and *The Fugitive* over and over. I had a small combo TV/VCR that I hauled into bed with me, and I'd watch the movies with my big toe on the power button. If sleep was near, I'd pop the TV off. I'd start the next night where I left off. I watched the movies in slices. If I finished one, I'd get halfway through another before my big toe twitched and sent the screen to black.

Amnesia. What it's not like is Geena Davis in *The Long Kiss Goodnight* chopping vegetables and then suddenly chopping faster and faster, having the family empty the crisper to keep up with her sudden ecstasy of chopping, shouting to the cutting board, "I know how to do this. I must have been a chef." And then throwing the knife, which sticks in the cabinet, and while it trembles, saying, "What? Chefs do that."

I found no secret talents.

For me it was more like Scott Bakula from the TV show *Quantum Leap,* who keeps waking up in other people's lives and figures out who he is by reacting benignly to whatever people say to him until he can get to a mirror and see who he looks like. Bakula always has a mission, to make something right, in order to placate the whims of some machine in the future that determines when and into what life he jumps next.

Amnesia is not like Guy Pearce in *Memento* waking up to find his leg wrapped in gauze, then peeling it off to find a tattoo with a cryptic command. I woke and found a tattoo on my ankle. All my tattoo told me was that I didn't like my tattoo.

Amnesia is a lot like being Arnold Schwarzenegger in *Total Recall* and dreaming of climbing the mountains of Mars with Rachel Ticotin, and then waking up in the arms of Sharon Stone, who acts like your wife for a while before she tries to kill you and you travel to Mars to talk to Kuato, who is a small man embedded in Marshall Bell's stomach.

No. That's a lie. It's not like that at all. But it is kind of like when Arnold sees a video of himself from the past, and the past Arnold is kind of an asshole mastermind whom the current Arnold hates. It is kind of like that.

It's not like Matt Damon in *The Bourne Identity* waking up in an ocean, either.

My friend Bob Trace—the guy who took a fistful of acid in Berlin, gave all of his possessions away to strangers, and was captured naked in the subway and taken to a German mental institution—came home to the States and took care of his dying grandfather. When I finally got around to calling him, he told me that this was his recovery process. He returned to the States and occupied himself with sponge baths and medication schedules (his grandfather's and his own). The house was in a rural and isolated area, and Bob went weeks without talking to anyone but his late-stage-dementia grandfather. He ended up watching *Forrest Gump* more than a hundred times. The movie became his touchstone, his two-hour-and-twenty-two-minute mantra. It was the handrail he clung to as he made his way back to mental health. I liked his technique but wanted to have a better movie than *Forrest Gump*, which is why I bought so many videos by auteur filmmakers. If I was going to be re-built by a movie, I wanted it to be a classic, preferably foreign.

There are some people who, when they have insomnia, can make productive use of those hours, reading or knitting as they wait for sleep to show up. My insomnia is not that kind. It's blurry, and my mind spins like a roulette wheel. I'm half-asleep/half-wild. I can't read. But I was overly ambitious when I was on my knees going through the boxes of two-dollar cast-offs. Movies with subtitles became just more reading. I'd get ten minutes into Kurosawa's *Drunken Angel* or Fellini's *Fred and Ginger* before I'd eject them and put in *The Fugitive*. Harrison Ford as Dr. Richard Kimble getting chased by Deputy Sam Gerard, played by Tommy Lee Jones—I didn't need any higher brain functions to track that plot.

Kimble's wife is murdered by a one-armed man, but the cops arrest Kimble for killing her, and he's sentenced to death. He escapes and goes back to Chicago to clear his name. He's lost everything that meant something to him, but he's going to continue to fight. For a vascular surgeon, Kimble is pretty streetwise. Kimble is a bruise of a man, but he continues his pursuit of the truth. He suffers; he endures. He's Job.

Sam Gerard is the stern US Marshal who chases after Kimble. He has one expression throughout the movie: steely exasperation. He's a man whose sole emotional reaction to the world is frustration. He's Eeyore.

Kimble chasing after an old life and Gerard chasing after Kimble: Job versus Eeyore. If you take both of these characters and cram them into your brain at once, then you have what recovering from amnesia feels like.

Watching a chunk of the movie until I fell asleep and then watching the remainder the next night made it seem like Kimble was always recovering his life, then losing it again. Kimble was shaking hands with Sam Gerard at the end of the film, and then moments later he was holding a gun on him inside of a dam. Kimble was confronting his nemesis in the banquet room of the Palmer House Hilton, accusing him of killing his wife, and then the movie would rattle and buzz and rewind, and there was Kimble, walking through a charity ball in his tuxedo, saving his beautiful wife from all of the other fawning doctors ("I was down to my last joke," she says, thanking him). Watching the movie this way created a fugue state for Kimble. He was guilty and innocent and jailed and freed and escaped and caught and sentenced and vindicated all at the same time. The story didn't matter anymore.

With a twitch of my big toe, it'd all start again.

The last time I saw my shrink I made a point not to be smart or clever or knowledgeable about my situation. I wanted to show her that I could learn, that I was cooperative, that I was willing to try it her way. That I was reachable.

I told her about my dreams. I told her about the banal-mares, in which I woke up as other people who were on the clock and waiting for their shifts to end. I told her about waking up as a convenience store worker and the chunk of me that was missing in that dream, and how it made me realize how crazy random it was not just that I was me (the billion sperm to one egg; the insane odds against all my ancestors ever meeting each other), but also how, with the millions of electrical pulses in the brain that were needed to fire every microsecond, it was incredibly random that I continued to be me. I yanked tissues from the box and pushed them into my eyes. I told her that I was tired of having my biggest fear be not being me.

"You know, convenience store workers serve a very important function in the universe," the shrink said.

"I know that," I said, thinking that she was missing something. "People need gas. People need cigarettes."

She leaned in close to me and said, "Convenience store workers are the reincarnated souls of people who died in the Holocaust. They need the mundane nature of those jobs to make sense of what happened to them in their previous lives."

"Oh," I said.

And just like that, I wasn't the craziest person in the room anymore.

I hadn't realized how tiring it had been to always suspect that I was the craziest person in the room no matter where I went. I stared into her eyes and was relieved of that burden. She was nuts. I recognized that. And because of this recognition, it was the first time I'd felt healthy in a year and a half.

Even so, the next week I called and canceled our upcoming appointments.

It was spring. I could tell by the progress of the construction
across the street. The stick frames of the new houses stood naked
in the wind. Men clambered around them, banging with hammers. I woke up most days having breathing problems: panic,
cigarettes, and asthma—my chest was clogged. I'd spend the
first half hour of every morning coughing myself clear. It was
a Saturday, and the prospect of an unplanned day stretched out
before me. I let Sally out and stood in my yard watching her
search for things to eat. The men were working across the street
even though it was a weekend. They were rushing to get as
much done as they could before the summer rolled in and baked
us all. My chest still felt tight, so I took a hit off my inhaler and
then lit up a cigarette.

For a week now, the windows of my house had been covered
with giant blackflies. I called the landlord and told him that
something had died in the walls, and he needed to take care
of it.

He asked me if I could smell anything.

I told him no.

He told me it was probably because of my dog and that it was
my job.

I told him Sally was exceptionally clean, but I knew this
wouldn't convince him and that I was never going to see him
actually do work at the house.

The thumbtack-sized black bodies danced on the windows,
hundreds of them. The night before I'd gone on a rampage with
an industrial-grade flyswatter I'd picked up at Walmart. After a
half hour the sills of every window were filled with dead flies,
but they kept coming, boiling out of the drywall. There was
something rotting inside the house, some rancid thing generat-
ing these bastards. I'd slept with the covers over my head and
tucked tightly around me. I woke up confused, sweaty, and de-

pressed, the damp cotton of my thin sheets plastered against my face.

I pulled the lawn chair up and decided to sit outside for a while. I could still see the tops of the mountains, but to see the lower parts I had to peek in between the two-by-four studs. Every day the construction took a little bit more of my view away.

I called my parents. My dad answered.

I told him what the shrink said.

He laughed. He asked if I was going to start seeing anyone else.

"I don't see what the point is," I said. "I'm not getting any better, and I'm just exhausting everyone around me."

My dad said that I wasn't exhausting him and that he knew my friends cared a lot about me. He told me that I had more people in my corner than I knew, that people really—

I cut him short. "I'm draining people. It's like instead of having a personality, I'm a virus. I clamp onto people and leech off them. I just need and need and need. I've ruined every relationship I've been in."

This is the way my depression works. When I'm in it, I'm always trying to define it, as if naming it correctly will make it disappear. But in the same way that knowing water's chemical composition doesn't help you to swim, this casting out of definitions does nothing.

"David. You're not a virus. You're an incredible young man who remembers things wrong. Do you remember when you were doing street performance with your friend from college?"

"The fortune-telling stuff?" My friend Duncan and I had lined a box with black velvet and put a baby doll's head inside of it. From behind the box we could manipulate the doll's mouth so it looked like he was talking. There was a switch that made him light up and a bulb that made him spit out a stream of reddish liquid. We called him the Child of the Apocalypse, and people would pay us a dollar for one of his prophecies.

"You called me after the first time you did it, and you'd made enough money to go drinking on; you had a big crowd of people, strangers, and they were all laughing at you and your friend. You went out again the next week—"

"And the guy spit on me." I lit up a new cigarette.

"Some old asshole spat on you."

"He was at a restaurant, sitting outside, and we were loud, and he said we ruined his meal and yelled at us. And when we left, he spit on me."

"Right," my dad said.

"This isn't helping," I said.

Sally came over and stood four feet away from me, her ears cocked forward and alert. She smelled something in the wind, and her whole body leaned toward the source.

"You're only remembering the guy spitting on you, Dave." Dad sighed. "All those other people didn't spit on you. And you forget about them being part of the story."

Sally started barking like crazy, and there were voices. Voices very nearby. But there was no one around. I went up to the gate, peering up and down the road: no one. The voices were still there and getting closer. Was I hallucinating again? Was Sally hallucinating, too? I told my dad I'd call him back. I circled the house looking for the voices. Nothing.

When I looked up, there was a wicker basket hovering right above me, and above it was a massive striped bubble. How long does it take the brain to recognize and name things? There are two sets of connections that leave the eye, and they're routed through different parts of the brain. One connection recognizes shape and line and shadow. The other connection names things. The first connection is a millionth of a second faster in a normally functioning brain. So for just a fraction of a moment I saw "brown square underneath large striped circle" before my other connections did their job and said, "hot-air balloon."

It was massive and hung just fifty feet above me. I could make out the edge of an elbow or a slice of haircut here and there as the passengers shifted about, gabbing. It was making a slow descent into the empty housing development across the street. I put Sally on the leash, and we crossed the street and watched the thing land, watched the people disembark, and watched it slowly collapse. The handlers arrived in a minivan with the company's name in neon yellow across the side. They gathered the balloon as it fell, grabbing fistfuls of the fabric and pulling it toward them, folding that riot of color against their bodies, then reaching out and grabbing more. They were sweating. This was real work.

I sat down with my back against the inner wall of the development, and I didn't want a truck to lose control on the highway and smash into the wall, leaving me bleeding and concussed until the ambulance came and spun around the cul-de-sac, pulling into the nearest driveway, collecting me and putting me on a stretcher, tying me down, and letting other people make all of my decisions for me.

No. I wanted to stay and watch this instead.

Things warmed up slowly between Anne and I. She had
stopped lugging around her giant mug of water and was only
swimming every other day. I heard from friends that she was
eating almost normally. It was right before spring break, and
the writing center was empty. We ran off extra evaluations and
sharpened pencils.

Clark didn't drop by. I felt like asking Anne how things were
going with him, but didn't. I'd never be someone she could
talk to about relationships. I'd heard that they were going to
spend spring break together, so maybe he was getting his car
checked out, maybe he was doing the billion little sweetheart
errands that a person would do to spend a week in the company
of Anne. He had stopped drinking as heavily. She was carrying
fruit around with her to munch on. I was becoming an extra in
the movie of their love.

As we busied ourselves with chores, the room was thick with
the conversations we should have been having. I needed to tell
her about Halloween and the woman with the cat ears. I wanted
to ask what she thought about the gifts of the scarf and the
childhood book I'd received. I needed to know if I had respected
her when we dated, and if I hadn't, I needed to apologize. I
wanted to know what our relationship had been like, what I had
been like. I never got up the nerve to ask any of these things.
The words that were shuffling around in my head wouldn't fall
into an adequate order for me to speak them. The shift ended.

Over the weekend, there'd been an assault in a parking lot on
campus, so I walked Anne out to her car. She didn't have a parking sticker, so she parked across the street in a church's lot. In
Las Cruces, the parking lots are the size of football fields. It was
spring, and the winds had come barreling down off the mountains, sandblasting everything with tiny detritus. We walked
sideways with our backs to the wind and were talking about the

English building, so named because no one who had graduated from it had enough money to rename the building after himself. We mocked the furniture, the paintings in the hallways, the lousy nubby carpet, the terrible lighting.

"It is the single most unerotic building ever constructed," I said.

Anne laughed and said, "We had sex in that building."

"You're lying."

She went on to tell me that my sister Katie had been in town visiting, and Anne had had a friend staying at her place. So we met at the English building and had sex in her office.

As she told the story, I remembered, not the day she was talking about, but the day that I first asked her out. I remembered her in a gray sweater, V-neck, and she was wearing glasses. She looked up at me over her glasses, and it was like someone had twisted my stomach around. I remembered wanting to be with her because of how she had looked up over her glasses at me. I remembered what that had felt like. It hadn't been lightning or thunder or fireworks or even a V-8 engine. It had been breathtakingly pleasant. I missed her. I was walking beside her and I missed her, missed us, missed that pleasantness. I had some of the memories, but they were static images. I had the where, the why, and the what she was wearing, but I didn't have the emotion. I knew the logistics of our relationship, but everything else was vacuum-sealed in plastic, stored somewhere I couldn't access.

We continued to walk sideways, sheltering ourselves from the gusts. She folded herself into her burgundy Civic and drove off, and at that moment I could finally make out the shape of what I had lost. And that thing that I'd been feeling, that lump of absence about the size of an apple that sat underneath my rib cage and pressed onto my stomach, that thing, while it didn't disappear at that moment, it did show signs of making the slow transformation from chest crushing to the slow pressure that Sally exerted when she leaned into me when I petted her. This pressure of absence and the anxiety that surrounded it would never leave me, but perhaps it could become just a quirk of my breed. My new breed.

In the story of my experience with Lariam I am always quick to point out how unlucky I was. Even though studies dating from 2002 say that 25 percent of people who take the drug have adverse psychological side effects, only a slim minority of that percentage reacts to the drug like I did. The specificity of my biology interacted with that drug and deleted whole sections of my life, leaving me alone and bewildered on a train platform thousands of miles from home.

When my mom tells my story, she always points out how lucky I am to have made it through alive.

After Anne drove off, the wind picked up more, and my jacket flapped around me, bits of flying debris pecking at it. The empty hooks on the flagpoles clanged and clanged. As it was dusk on the Friday before spring break, the parking lot was deserted, just yellow lines on cracked asphalt limning where things should go with a faith that someday they'd be filled up again.

I was alone, but I knew where I was going because Sally was at home and needed to be let out.

POSTLUDE

I know this isn't much.

But I wanted to explain this life to you, even if

I had to become, over the years, someone else to do it.

—Larry Levis, "My Story in a Late Style of Fire"

Table 1. Various clinical presentations of neuropsychiatric effects of mefloquine. Exhaustive listing of reported adverse CNS reactions observed in people who received mefloquine.

Major psychiatric disorders and symptoms	Delirium, delusion, hallucinations, illusions, megalomania, paranoia, psychosis, schizophrenia
Disorders of affect	Aggression, behavior disturbance, character change, depersonalization, depression, euphoria, hypomania, logorrhoea, mania, mood swings, oppression, personality disorder, suicidal, suicide attempt
Neurosis	Hypochondriasis, malaise, mutism
Other psychiatric symptoms	Abnormal hunger, agitation, aggravation, amnesia, *angor mortis* (the feeling of imminent death), anorexia, anxiety, apathy, asthenia, disturbed awareness, reduced concentration, confusion, dazed, disorientation, dreams, drunken state, excitation, exhaustion, fatigue, fear, hyperventilation, insomnia, memory impairment, nervous, nightmares, panic reaction, restless, somnolence, speech disturbance, sweating, tiredness, decreased alertness, vegetative dystonia, weakness
Seizures	Aggravated seizure, convulsion, clonic seizure, epileptic seizure, epileptiform fits, generalized seizure, grand mal epilepsy, fits, tonic-clonic seizure
Disturbances in level of consciousness	Acute brain syndrome, cerebral edema, cerebral ischemia, clouded consciousness, coma, encephalopathy, encephalomyelitis, obnubilation, semiconscious, stupor, unconscious
Dizziness	Abnormal coordination, ataxia, balance disorder, dizziness, unsteady gait, lightheaded, loss of balance, uncoordination, vertigo, walking difficulties

Neuropathies	Anomia, cranial nerve disorder, abnormal EEG, twitching eyes, foot or hand paresthesia, general spasms, hearing disturbance, hypesthesia, leg paresis, leg pain, lip paresthesia, muscle weakness, myalgia, neurological disorder, neuropathy, numb fingers, numbness, paralysis, paresthesia, paresis, polyneuropathy, Raynaud's disease, sensory disorder, slow reactions, tinnitus, tongue spasm, vision disturbance, weakness
Headache	Aggravated migraine, cephalgia, eye pain, headache, head pressure, migraine
Other neurological disorders	Abdominal pain, back pain, chest discomfort, chest pain, cramps, dystonia, fall, fever, gastric colic, hot flushes, incontinence, intestinal spasm, limb pain, lumbago, muscle tremor, edematous legs, esophageal burning, oropharyngeal spasm, pallor, rigors, shakiness, shivering, stomach pain, tetany, thirst, tinnitus, trauma, trembling, tremor, twitching, visual disturbance
EEG = electroencephalogram	

(Adapted from Francois Nosten and Michele van Vugt, "Neuropsychiatric Adverse Effects of Mefloquine: What Do We Know and What Should We Do?," *CNS Drugs*, Volume 11 (1999): pages 1-8, Table 1, with kind permission from Springer Science+Business Media B.V.)

In late July 2007, I was living in Houston, Texas, and one night I had a series of seizures that lasted until morning. They were accompanied by minor visual and auditory hallucinations.

Clark and Anne were still together. They had moved to Portland and were teaching up there, making a real adult life of it. Veda had finished his dissertation, and he and I had stopped writing each other. His daughters were teenagers, and his life was much busier than it was when I knew him. The world had moved on, but I was still having side effects. I was unable, at a biochemical level, to get over it already.

My doctor sent me to a neurologist. I told the neurologist that years before I'd had an allergic reaction to Lariam.

"No, you didn't," he said.

I explained to him that I had elevated levels of immunoglobulin E in my blood right after the experience.

"That may be true," he said. "But what happened to you had nothing to do with those levels."

I had memorized immunoglobulin E and kept that information handy for all of these years. And it meant nothing. No one could tell me what had happened to me, though they could tell me what hadn't happened. It was maddening.

I got an MRI, another day spent undressing, removing all metal, and being slid into the tiny clanking tube, but when it ended up clean, I skipped the EEG and the three follow-up appointments. I didn't have the money for it all. I went to a writers' conference in New England instead. Because of the seizures I scored a single room. I met a woman there named Emily. She hated smoking, so I quit.

Back home, I made an appointment with a new therapist, who gave me sleeping pills for my insomnia. I bought a pair of running shoes, and when the traffic had died down and my mind was running at full carousel, I'd go out into the humid Houston night and run until my nipples bled.

The US military, after years of claiming that Lariam was perfectly safe, stopped prescribing the drug as its primary malarial prophylaxis for its troops as of October 2009. They still use it in special situations, though. As of spring 2012, the DOD acknowledged that Lariam was still being given to US soldiers stationed in sections of Afghanistan. On June 6, 2012, Dr. Remington Nevin, an Army epidemiologist, testified to the Defense subcommittee of the US Senate Committee on Appropriations that Lariam was the "Agent Orange of our generation" and that we will be seeing its effects on our veterans for years to come.

In 2008, the DOD released a slew of documents pertaining to suspicious suicides that took place at the prison in Guantanamo Bay. Deep within these documents was evidence that beginning in 2002, upon arrival each new inmate was given an initial dose of 1,250 mg of Lariam—a massive dose (five times the prophylactic weekly dose)—before they were tested as to whether or not they were infected with malaria.

Guantanamo Bay doesn't have malaria. Cuba doesn't have malaria. None of the soldiers or contractors working at the base were prescribed anything for malaria. Not to mention, why would you give someone a massive dose of any drug before you knew whether or not they had the disease you were treating them for? Especially if the drug you were administering had a track record of awful psychological side effects?

Professor Mark Denbeaux, director of the Seton Hall Law Center for Policy & Research and legal counsel to Guantanamo detainees, believes that since there is no sound medical reason for administering such a high dose to every detainee, then the psychological side effects were the intended primary goal. In what has been called "pharmaceutical waterboarding," the megadose of Lariam given to each detainee upon arrival, regardless of malarial status or of past medical history, was administered with the knowledge that the incidence of side effects goes up with the level of dosing. In the guise of preventative medicine, the military found an efficient way to psychologically terrorize prisoners and soften them up for interrogation.

In 2009, the woman I had met at the writers' conference in 2007, Emily, came to Texas to visit. We'd been dating on and off since we'd met. She and I were driving around in the hill country of west Texas, a mile or so south of a famous giant telescope. We were new enough in our relationship that car silence was still a novelty for us. Among that incredible undulating landscape, there we were, seeing if it was tenable not to always be talking. Emily was driving.

We came up over a little rise, and there in the opposite lane was a woman sitting in the road. We saw her first, then the thirty yards of scattered metal and plastic wreckage, and then saw another woman one hundred yards farther down the road, lying next to a motorcycle. A man up a ways was climbing off another motorcycle and running toward the second woman. Emily pulled the car to the side of the road and jumped out.

I'll tell you now that I fell in love with her at that moment. She jumped out. No dithering. I was turning to ask her what we should do, and she was already gone. She put the car in park and jumped out. I reached over, snapped off the ignition, and followed.

Emily went to the woman lying next to the motorcycle, and I went to the other woman. Her name was Alma. She looked about forty and was wearing high-cut jean shorts and a tank top. Most of the skin on her legs and shoulders was grated off, and her bare feet—she'd originally been wearing flip-flops—were swollen and blood soaked; her pedicure was chipped, and some of her foot bones were visible. Her boyfriend, Joe, was down a shallow ravine, sprawled next to their motorcycle and going in and out of consciousness.

What had happened was that Joe and Alma were out riding their motorcycle and were coming up the steep hill while the other couple, Philip and Gigi, were on their respective motorcycles coming down. Gigi took the turn too wide and strayed into Joe and Alma's lane, sideswiping them. Joe and Alma skidded across thirty to forty yards of asphalt. Gigi dumped down herself a hundred yards farther down the road. Everyone had been wearing helmets, so this isn't as terrible a story as it could have been.

Because we were first on the scene, Emily and I were in charge. Someone else had showed up and started to work on Joe. With no medical gloves, I tried to avoid any contact with Alma's blood. She'd been propping herself up on her elbows and I was tentatively smoothing her hair back when she collapsed into my arms.

I spent two hours holding this stranger. The accident had happened in a cellular dead zone. The nearest working phone was twenty minutes away, and the nearest ambulance was much farther than that. Her shoulders were a tangle of gravel and flesh. Alma's blood was on my hands, on my neck, in my hair — my favorite shirt was ruined by it. When the ambulance showed up and the paramedics put her on a stretcher and loaded her into the back, I was surprised not to be asked to go with her. The door was slammed shut, the sirens wailed, and she was gone. And I stood there as unthanked as any tourist police officer working the train station beat.

In the chaos of this world, where we carom and collide in the everyday turbulence, there's something about the specific gravity of the helpless individual, the lost and the fractured, that draws kindness from us, like venom from a wound.

Lariam is a lipophilic (fat-soluble) molecule and has a long half-life. If the drug accumulates in the brain, which is 60 percent fat, it can remain in the body for a long time before it is eliminated entirely. If something dislodges the Lariam molecule during this time, it could end up clogging another protein gap junction. Some people have experienced side effects ten years after their last dose. This is one hypothesis. The other is that the Lariam outright poisons the brain and the resulting damage is permanent.

My last dose was October 10, 2002.

As I write this in September 2012, I am still afraid of lingering chemical instabilities; afraid of what might still be stuck in the wrinkles and folds of my gray matter; afraid of what might get dislodged, disrupting who knows what electrical signals; afraid of where I'll wake up next.

At what point is this new language—*protein gap junction, half-life, lipophilic*—another kind of superstition, a peer-reviewed set of erudite notions that I've used to chase my boogeymen away? I started clutching at science, as if a specific enough diagnosis would work like a flashlight shone underneath a bed to prove there are no monsters. But even a child knows that any monster worth his fangs will hide until the lights are turned off again.

Octobers vexed me for years. The anxiety had metastasized from the container of the single anniversary day of the seventeenth to infect the entire month. I'd call in sick, stay drunk, and chain-smoke until Halloween showed up.

But on October 1, 2011, I married Emily, the woman who jumps from cars to help people, and her birthday is in the middle of October. Now this month is full of dinners out and present exchanges. I'm finding ways to rebrand October, to celebrate the life that I have while I have it. Some days, I feel perfectly fine and amazed that anything terrible has ever happened to me. Other days, it seems irresponsible that I'm allowed to cross the street by myself.

But this, in comparison to what I've been through, is everyday crazy, and everyday crazy is something I can handle.

In chapter 4 of *The Great Gatsby,* there is a three-page section where the narrator, Nick Carraway, lists all of the people who visited Gatsby's house during that summer. It's one of my favorite parts of that book. This list of people, with their semifamiliar names and attached anecdotes, acts as the adjective to the noun of Gatsby, the noun James Gatz, the poor kid from North Dakota, is trying so hard to invent in order to impress a woman. As these people wheel around at his marvelous parties, Gatsby is allowed to be the empty spoke.

There are beautiful people, wealthy people, fabulous people, artistic people at the party; the attendees confer shards of those identities to the host. The guests know nothing about Gatsby. He's just a guy who arrived in West Egg and started throwing extravagant parties. At the same time, the guests know everything about Gatsby. They all have excellent information from impeccable sources.

> *"He's a bootlegger," said the young ladies, moving somewhere between his cocktails and his flowers. "One time he killed a man who had found out that he was nephew to Von Hindenburg and second cousin to the devil."*

Those days in the Indian asylum where I was fed heavy meds, I was a lot like Gatsby, except I wasn't aware or in control of who was invited to the party. People treated me a certain way, and I became the kind of person who was treated like that. All I had to go on for my identity was the reactions of the people arrayed around me. I assembled a working self out of the behavior of others.

Rajesh/Josh, the cop, told me I was a drug addict; I became a drug addict.

Mrs. Lee told me I was killing my mother with my behavior; I expressed regret for my awful actions.

Richard offered me cigarettes; I became a smoker.

My parents told me I was a Fulbright scholar; I tried to act smart.

Anne said she loved me; I told her I loved her back.

I allowed the people who showed up to define me. I didn't have much choice, really.

Those days in New Mexico I was also a lot like Nick Carraway. I was constantly hearing about this guy named David. People told me stories about him. I was always trying to locate him in the crowd of strangers. I'd act like him in order to bring him closer to me.

I found my old comic books and made those stories tell me something about myself.

I studied old pictures of myself and mimicked the poses.

I found my books with marginalia written in my handwriting and sifted through them until I discovered what I had found so interesting in the first place.

I used the tools that were at hand: the possessions stored in my parents' attic, the stuff piled in a storage locker in New Mexico, the entire albums of group photos I'd ruined by making faces.

I made myself wear a mask that resembled this other guy, and it amazed me when it fooled everyone.

Nick spends the party trying to find Gatsby so that he can thank him for the invitation. He and Jordan Baker tour the house looking for the host. Behind an "important-looking door" they find a lavish library, and in the center of it is a middle-aged "owl-eyed" man. Though he looks the part, large and alone in a room full of things, he is not Gatsby. This man is a drunk guest who is marveling over how well everything is put together. He had expected the books in the library to be fakes, but he holds one up in surprise and says, "It's a bona fide piece of printed matter."

Nick and Miss Baker exit and continue their hunt for Gatsby.

At the end of the night, Nick is sitting next to a man. The man says that Nick looks familiar, that he recognizes him from the war. The two men reminisce for a bit about "some wet, gray little villages in France." Nick mentions to the man that he's frustrated about not having met the host. You know the story. This anonymous man Nick found himself sitting beside and chatting with—it's him. It's Gatsby.

You fight and search and fight and search, and when you give up, you find yourself sitting next to what you've been looking for. And it recognizes you before you recognize it.

This is a long way of going about saying that I was, of course, sitting right next to him the whole time. Me, I mean. I was sitting right next to me the whole time. Even if I'm as much an invention of myself as Gatsby is of poor James Gatz, there are some bedrock fundamentals: I have a talent with children (Amol at Woodlands mental hospital, the three boys on the moped); I'm a good teacher (good enough for a student to contact me a year later about his cancerous genitalia); I am frequently conflicted about any and all intimate relationships in my life; I can be funny (I was the most entertaining psychotic at Woodlands), and I can also use humor to keep people at a distance; I have worn absurd costumes and have a fondness for saying absurd things; I have a long record of screwing up group portraits; I have parents and two sisters who love me and whom I've loved back as best as I am able.

I am lucky.

Like Nick Carraway, I walked through the cacophonous disorienting party looking for the host, and at the end I was sitting right next to him. To me. I finally discovered this guy who was me.

288   After all of the talk and stories and lies, I'd never have recognized him.

ACKNOWLEDGMENTS

This book owes a strong debt to Kathleen Lee. She is a great friend and encourager. Honestly, no one in the world is as much fun to be miserable with as Kath.

I was just a twenty-year-old know-nothing when I stumbled into a lecture by Robert Boswell. He has become, in the years since that night, a great mentor, teacher, drinking companion, and friend. The first time I met his wife, Antonya Nelson, it was ten a.m., I was fixing her toilet, and she told me the only polite thing to do was to pour us both drinks. She's been keeping me laughing (and drinking) since and has challenged me in a million ways to become a better writer, reader, and instructor. I can't thank these two enough for their warmth, their integrity, their brilliance.

Tony Hoagland taught me more about writing sentences than I ever thought possible.

I thank my agent, Eleanor Jackson. She's funny and she likes running, whiskey, and literature. There is no better combination of traits possible.

I thank my editor, Lauren Wein. As soon as she and I sat down over lunch, I knew I had found the perfect person to help shepherd this crazy book through draft after draft.

I thank Sarah Koenig, Nancy Updike, Ira Glass, and everyone at *This American Life* for featuring my essay on their show.

Ladette Randolph runs *Ploughshares,* and her tenacious commitment to literature is something rare to behold. I count myself blessed to have been featured in that magazine.

Big thanks go to the PEN American Center for all the great work they do.

I thank Warren Wilson College, New Mexico State, the Fulbright Foundation, and the University of Houston for educating me and for introducing me to the finest colleagues a person could ever want.

290   The United States Education Foundation in India has some amazing people working there. Jane Schukoske and S. K. Bharti were always kind, patient, and helpful during the worst of the worst.

The Inprint organization of Houston helped me survive graduate school with several grants and teaching opportunities. They are a great group of people who do wonderful work.

I tell a couple of stories about terrible counselors in this book, but it was a great one, Dr. Allison Godby, who helped me to begin to tell this story without hyperventilating.

I thank Scott Repass and Dawn Callaway for being great friends. Scott and Dawn make me laugh, think, and work harder. They also make me drink better. One of the great things I get to do in my life is to visit Houston and drive around with the two of them making jokes. Or as R___ would say, "Ay-ohh!"

Thank you to Jamie Thomas, Noah Boswell, Alex Parsons, Mat Johnson, Eric Beverley, Jenn Stennis, Miah Arnold, Casey Fleming, Greg Oaks, Peter Hyland, Bradford Telford, Aaron Reynolds, Jill Stukenberg, Peggy Chapman, Connie Voisine, Rus Bradburd, Casey Gray, j. Kastley, Kathy Smathers, Jon Lillie, Laura Zebehazy, Karen Oh, Chris Rado, Miriam Carillo, Melissa Lillie, Tim Rice, Scottbutt, Sam Scoville (without whom nothing would've ever happened ever), Lillie Robertson, and Rich Levy.

Duncan Trussell and Emil Amos I thank for laying the groundwork for my fascination with India by traveling the country with me back in 1998. They're also two of the wildest and most consistently original brains on the planet.

At one point in my writing life, I needed a kick in the ass. The generous Sarina Pasha was more than happy to provide a literal one during her happy hour shift at the Poison Girl bar in Houston.

Thanks go out to the Schnaars, the Hustons, the Runyons, and the Webers. As well as to all of my wife's Chicago friends who have welcomed me and have become my friends: Mike Hawthorne, Paula Tordella, Lauren and Adam Nevens, and Josh Noel. And thanks to my new sisters (they came with the marriage), Deb and Madeline Stone.

I thank Jeanne Lese of Mefloquine Action (www.lariaminfo .org), a great virtual clearinghouse for Lariam information and a meeting place for those afflicted by the drug. For anyone looking for more information about Lariam, her website is a great starting place. I thank Dan Olmstead and Mark Benjamin for being dogged reporters. Their articles on Lariam and its effects are informative and harrowing. Sonia Shah's book, *The Fever: How Malaria Has Ruled Humankind for 500,000 Years* is a great read on mankind's long, long history with malaria. And Andrew Spielman and Michael D'Antonio's book *Mosquito: The Story of Man's Deadliest Foe* is also outstanding and a great read.

I thank my trio of scientists, Dr. Cindy Voisine, Dr. Rob Mitchum, and Lisa Jarvis, who helped me grasp the science documented in this book. And I am very grateful for the help of Dr. Remington Nevin, who gave the book a last minute fact-check.

I thank Jan O'Callaghan, who gave me her permission to reprint her son John's suicide note.

I thank Errol and Susan Stone for letting me use their cabin as a writing retreat and for being excellent in-laws.

Big thanks to Veda, without whom I would not have survived.

Thank you to Rayna Gellert (the best damned fiddle player in America and whose album *Old Light: Songs from My Childhood & Other Gone Worlds* has a song she wrote inspired by my experience.); Brett Nolan (a man who's never afraid to talk about the amount of time he spends spooning his cat); Gillian Flynn (Baguette Dork's favorite aunt); and Shale Aaron and Stu Spears. An amazing and patient group of early readers to whom I owe much.

ACKNOWLEDGMENTS

Thanks to Niko and Ezio—astonishingly good nephews.

Thanks to Lydia, my beautiful brand new daughter. Someday I'll give you a full explanation of St. Frankenstein, if you'll explain to me why Devo is one of the only bands that gets you to sleep.

Thanks to my family. I woke up to a life in progress and found that I was lucky enough to have you all in my life. Mom, Dad, Katie, Betsy, Eric, Candymom—I love you all and am lucky enough to know how lucky I am to have you in my corner.

And finally, thank you to my wife, Emily Stone. I'm looking at the back of her head as I write this, and it's Saturday morning, and I can't think of anything more fortunate than to get to spend an entire Saturday with her.

Chicago, Illinois
April 2013

292